THE STORY OF ELECTRICITY

THE STORY OF ELECTRICITY

With 20 Easy-to-Perform Experiments

George deLucenay Leon

DOVER PUBLICATIONS, INC., NEW YORK

Drawings by Man-Kwan Chiang
based on sketches provided by the author.

Published in Canada by General Publishing Company, Ltd., 30 Lesmill Road, Don Mills, Toronto, Ontario.
Published in the United Kingdom by Constable and Company, Ltd.

This Dover edition, first published in 1988, is an unabridged, slightly corrected republication of the work originally published in 1983 by Arco Publishing, Inc., New York, under the title *The Electricity Story: 2500 Years of Experiments and Discoveries.*

Manufactured in the United States of America
Dover Publications, Inc., 31 East 2nd Street, Mineola, N.Y. 11501

Library of Congress Cataloging-in-Publication Data

Leon, George deLucenay.
[Electricity story]
The story of electricity : with 20 easy-to-perform experiments / George deLucenay Leon.
p. cm.
Reprint. Originally published: The electricity story. New York : Arco Pub., 1983.
Bibliography: p.
Includes index.
Summary: A history of electricity over the past 2500 years with twenty safe and simple experiments demonstrating some of the most vital discoveries.
ISBN 0-486-25581-6 (pbk.)
1. Electricity—History—Juvenile literature. 2. Electricity—Experiments—Juvenile literature. [1. Electricity—History. 2. Electricity—Experiments. 3. Experiments.] I. Title.
QC527.2.L46 1988
537'.09—dc19 87-27571
CIP
AC

To the inventors and scientists
of future generations:
May this book prove to be a source
of inspiration and knowledge
to those destined to advance
magnetism, electricity, and electronics
to yet undreamed-of goals

Contents

Acknowledgments

Thanks to:

Consolidated Edison, for providing rare photos of Thomas Edison and of his inventions, as well as a free run through their museum to photograph exhibits;

Heath Company, for permission to reproduce schematics from their manual;

Radio Electronics magazine, for permitting me to copy a schematic that originally appeared in that periodical;

Radio Shack, for giving me a kit to build the Motor/Generator;

Charles Ruch of Westinghouse Corp., for information about Tesla's relations with George Westinghouse.

Faraday's electric generator and self-induction drawing are from his own laboratory notes and reproduced through the courtesy of Harper & Row.

List of Materials for Experiments in This Book

Amber
Plastic or glass rod
Paper
Thread
Dry grass
Feather
Needle
Horseshoe or bar magnet
Small piece of cork or wood
Polyethylene (Saran Wrap, for example)
Pencils
Iron filings (from steel wool pads)
Piece of wool or fur
Wire—one spool of size #22
Supports and holders (see Experiment No. 6)
Silk thread
Glass tube 2 feet long
Nail
Wide-mouthed jar or bottle
Aluminum foil
Wooden handle glued to tin can
Phonograph record
Paper towels
Rubber glove
Ten pennies and ten dimes
Vinegar
Meter—0 to 50 milliamps
Iron or steel wire
Lemon
Copper wire
Compass
Enamelled magnet wire
"B" or "C" battery
Cardboard sheet
Cardboard roll
Darning needle
Six-volt lantern battery
One 1.5-volt lamp
Two 1.5-volt cells
On-off switch
Heavy insulated wire
Iron bar or large nail
Candle
Large copper disk
Crank
Ten-ohm resistor (5 percent or better tolerance), 2 watts

Sources for these items are 5-and-10-cent stores, supermarkets, hardware stores, and electronic supply houses like Radio Shack and Lafayette Radio.

You can order various kinds of electrical equipment, such as meters, magnets, generators, kits, and solar cells, from a number of mail-order suppliers. Write to the following companies for a free catalog:

Carolina Biological Supply Co., 2700 York Rd., Burlington, NC 27215
Delta Education, Inc., P.O. Box M, Nashua, NH 03061-6012
Edmund Scientific, 101 E. Gloucester Pike, Barrington, NJ 08007
Fisher Scientific Co., 4901 W. Le Moyne St., Chicago, IL 60651
Frey Scientific Co., 905 Hickory Lane, Mansfield, OH 44905
LaPine Scientific Co., 6001 S. Knox Ave., Chicago, IL 60629-5496
McKilligan Supply Corp., 435 Main St., Johnson City, NY 13790
NASCO, 901 Janesville Ave., Fort Atkinson, WI 53538
Radio Shack, P.O. Box 2625, Fort Worth, TX 76113
Sargent-Welch Scientific Co., 7300 N. Linder Ave., Skokie, IL 60077
Science Kit, Inc., 777 E. Park Dr., Tonawanda, NY 14150
Sheldon Laboratory Systems, P.O. Box 1059, Jackson, MS 39205

When you get one of these catalogs, you will find a pleasant surprise: lists of materials and equipment used in other sciences—like biology, geology, astronomy, and chemistry. Books about science and books of experiments also are listed.

Preface

This book has two purposes. The first is to trace the history of electricity. The second is to introduce some of the experiments that played a vital part in this history.

Most of the objects employed in the experiments can be found in the home. A few things need to be bought. None are expensive, and *no experiment is dangerous.* A six-volt lantern battery powers those experiments that require electricity.

All the experimenters and inventors who contributed to our knowledge of electricity could not be included in a book of this size, so I put those in who best illustrated a specific point. Those inventors who were better known for their work in another science were generally left out, with the exception of two giants whose ranges of interest were all-encompassing—Edison and Tesla. If you know of others who should have been included, I leave it up to you to study them yourself.

Meanwhile, enjoy the experiments.

Introduction

People are naturally curious. Most of us want to know what makes things work or why natural forces behave as they do. What makes lightning? Hurricanes? The rain? Humanity has been asking these questions for centuries.

Of course many animals are curious. A blackbird will pick up something shiny and hide it in its nest. A dog will sniff out an unfamiliar odor. However, there is one major difference between the curiosity of humans and animals.

The difference is that we not only question the things around us: we put them to work in some useful way. From simple curiosity about the origin of lightning, we have developed the sciences of magnetism, electricity, and electronics.

Through magnetism, the compass made worldwide exploration possible. The ability to generate large quantities of electricity has given us a very comfortable way of life. International communications, medical diagnosis, lights in our homes, heating and cooling—all come from experiments that began with the study of the effects of magnetism and electricity.

Today, the science and technology of electronics makes computers possible. These machines make complex calculations—which used to require hours of figuring—within fractions of seconds. It was the giant step we took from magnetism to electricity, and its off-shoot, electronics, that enabled us to launch space vehicles that probe distant planets and send us back pictures of their moons and rings.

So we can truthfully say that studying the effects of magnetism eventually took us to the stars. In this book, we're going to trace some of those steps. Come join me as we travel across many lands and over 2,500 years of history. We are going to meet scientists from many nations and cultures who used their curiosity to discover the secrets of electricity. The trip promises to be an exciting one.

THE STORY OF ELECTRICITY

1

A Static Attraction

Early humans were astounded by four distinct wonders that we now know to be electrical:

1. Lightning.
2. The way amber attracted light objects.
3. "St. Elmo's Fire"—a glow or a flame sometimes seen on the tips of masts during stormy weather. (St. Elmo was the patron saint of sailors.)
4. The way certain fishes, such as the electric eel and the torpedo-fish, stunned their prey.

Originally, many of these phenomena were explained by myths. For example, the Vikings believed that Thor, the god of thunder, owned a magic hammer. He would hurl it down to earth, creating lightning as it spun down. The Dutch settlers up the Hudson River invented a humorous folk tale in which thunder was caused by the gods of the nearby hills who were knocking down pins in a bowling game.

Not until the eighteenth century was it realized that lightning was indeed electricity—millions of volts being discharged from one cloud to another or from a cloud to earth.

Electricity began to be recognized for what it was less than 300 years ago. Magnetism, on the other hand—or what was thought to be magnetism—has been studied for about 2,600 years. It was this study that led to the discovery of how to generate electricity.

Electricity does not naturally exist in quantities that we can control and use for our benefit. There is no rock, no liquid, no kind of air that gives us electricity directly. We can generate electricity from the sun, from waterfalls, from coal and oil, but only after many intermediary steps.

Over the centuries, many men experimented with electricity and magnetism. They developed theories and tested them with experiments. From the results, they could tell if they were on the right track. They would then develop further theories and make another little stride along the road of knowledge.

Using this book, you will be able to understand these men's thoughts by repeating their experiments.

(Weren't there any women studying magnetism and electricity? Unfortunately, for centuries women were not allowed to study science. It was considered exclusively a man's subject. It was not until 1911, when Marie Curie won a Nobel prize in chemistry, that women came to be recognized as significant contributors in the area of science.)

Thales "Discovers" Static Electricity

To begin with, we must travel backward in time to Greece in the seventh century B.C.—almost 2,600 years ago. The place is Miletus, an ancient city on the west coast of Asia Minor, where a famous philosopher and mathematician named Thales lived. He lived from 640 to 546 B.C. and was the first person to have recorded interesting things about static electricity and magnetism. There were probably people before Thales who were aware of static

EXPERIMENT NO. 1: "MAGNETIZING" AMBER

Materials:
Amber, or plastic or glass rod
Piece of paper, length of thread, dry grass, feather
Needle

If someone in your family owns an amber necklace or bracelet, ask to borrow it. You will not damage it. If you can't get amber, you can do the experiments with a glass or plastic rod. Rub the amber until it becomes warm. Then bring light objects near it. (This will work best on a cold, dry day.) Do your experiments check with Thales'? You should be able to lift up a tiny piece of paper with the electrically charged amber. After a few seconds, it will fall off. Rub the amber on a piece of silk or nylon. Is there any noticeable difference in the amount of attraction from when you rubbed the amber on a piece of wool?

Rub the amber. Bring it close to the steel needle. Does anything happen? Experiment on your own with other objects, making a list of which ones are attracted to the amber. Are any repelled?

electricity and magnetism as curiosities, but they left no writings to tell us about their experiments.)

Thales found that amber will attract light objects such as feathers, bits of dried grass, and straw. We can only guess at the events that led to his discovery. One day, Thales was perhaps polishing some amber with a piece of fur or wool. To his surprise, when he laid the stone near a straw, it jumped and clung to the amber. Curious, he wanted to see what amber would do to other light objects. Feathers and dozens of other objects were tried. The results were the same. The light objects fastened themselves onto the amber.

He repeated the experiment, but this time without rubbing the amber. Nothing happened. Again he rubbed the amber, and again the feather and other objects responded. He must have repeated his experiments countless times to make certain that what he was observing was not just lucky chance. Eventually, Thales came to the conclusion that it was his rubbing that made amber "magnetic." (Actually, as you will see in awhile, rubbing the amber created a static electric charge and not magnetism.) You can duplicate Thales' experiments in Experiment No. 1.

Every good experimenter repeats his experiments several times. To be of value, the results must be the same each time.

Thales noticed that while rubbed amber attracted light objects, it did not attract metal. But when he picked up a piece of lodestone (an iron ore with a silvery finish, also called magnetite), he found he could attract pieces of iron. He also noted that lodestone attracts iron without being rubbed.

Later, other experimenters found that a variety of substances, such as diamonds, were able to attract the same way that amber did. Such substances we now call *insulators*. They do not conduct electricity. The rubber around an electric cord is one example of an insulator. It prevents electricity from escaping from the wire to anything near the wire. On the other hand, lots of substances do not attract metals, paper, feathers, or anything else, no matter how long you rub. We call these substances *conductors*. A conductor is able to allow electricity to flow through it. Examples of conductors are copper, silver, or gold wire.

Until the seventeenth century, it was believed that rubbed amber and magnets such as the lodestone had one thing in common—the ability to attract other objects. The explanation then was that magnets and amber did something to the air around them, and this was what caused the attraction. No difference was found between the action of the amber and the lodestone.

Static Electricity

Walk across a nylon or wool rug on a cold dry day. Now touch a doorknob, or any metallic object. You will get a slight shock, and sometimes you will hear a small crackling noise. If you have an AM radio nearby, you might hear the crackling sound amplified through the loudspeaker. If you do this experiment in a dark room, you will see a spark leap from the tip of your finger to the metal, like a tiny bolt of lightning. This type of electricity is called *static electricity; static* is derived from a Greek word meaning "standing."

The electricity is at rest and moves only from you to the metal you touched in one fast movement. Touch the metal again and nothing happens. You "discharged" yourself the first time. Walk across the rug and the discharge will take place once again.

Why? All matter is made up of atoms. The book you are holding, the chair you are sitting in and you, yourself, all are made up of atoms. The atom is the smallest particle into which an element can be divided and still have the chemical properties of that element. Some elements you may be familiar with are oxygen, hydrogen, iron, gold, carbon, and many others. So far, scientists have found 96 elements on our planet.

Atom is a Greek word meaning "indivisible." When atoms were first discovered, scientists believed nothing was smaller. Since we can't see an atom, it is easy to understand why they thought so.

An atom has a tiny, dense core having a *positive* electrical charge. Its weight accounts for almost 95 percent of the total weight of the atom. This core, or nucleus, is surrounded by one or more *negatively* charged particles called electrons (see Figure 1). The *negative* electrons and the *positive* nucleus balance each other. They remain that way until we scrape our feet across the rug or rub a piece of amber or plastic. At that moment, the material being rubbed picks up *negative* charges from the material used in the rubbing.

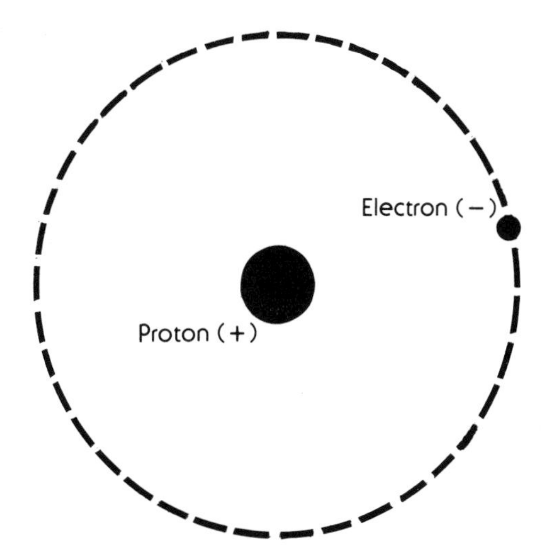

Figure 1. Schematic of a hydrogen atom, the simplest of all atoms, containing one proton and one electron. The figure is not drawn to scale.

When the atoms are out of balance there is a sudden, very tiny discharge that balances the atoms once again, as the rubbed object gives up *electrons* to the rubbing material. That is why you felt the small shock when you touched the doorknob.

Experiment No. 2 is going to prove something very important about static electricity. First, you will show that when two objects have like or similar charges they will repel each other. In other words, if they are both negative or are both positive, they will push away from one another. On the other hand, if the one is positive and the other negative, they will attract each other. This attraction will continue *only* until they have exchanged electrons and have the same charge. Then they will shove each other away.

EXPERIMENT NO. 2: GETTING A CHARGE OUT OF POLYETHYLENE AND WOOL

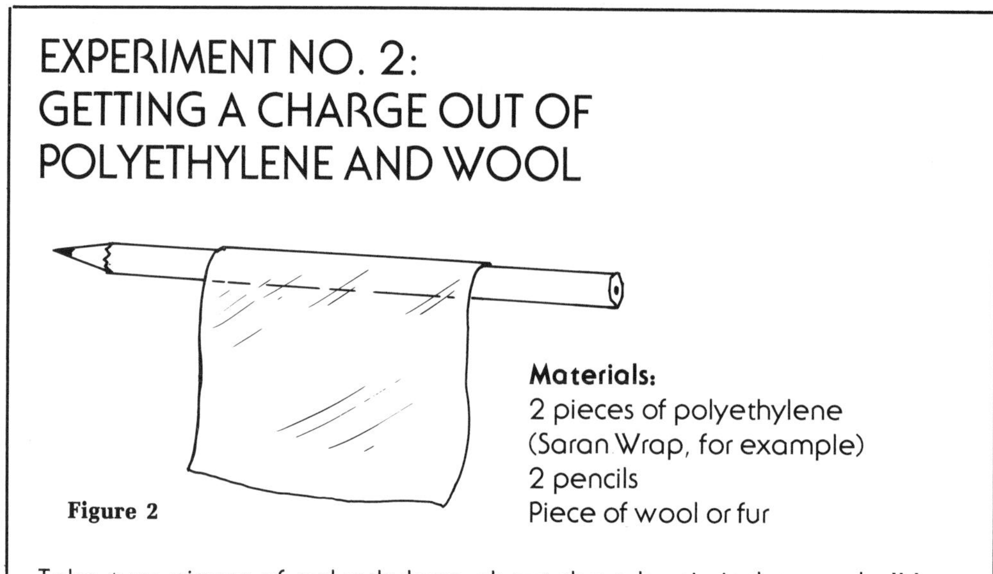

Figure 2

Materials:
2 pieces of polyethylene
(Saran Wrap, for example)
2 pencils
Piece of wool or fur

Take two pieces of polyethylene about three by six inches each. Wrap each of the poly sheets around a pencil so it hangs down like a flag (see Figure 2). Rub each sheet with your hands or against your clothing. Bring the "flags" close together so the sheets would normally touch. But they don't. On the contrary—they pull away from each other.

By now you know why they behave that way. According to atomic theory, when you rubbed the pieces of poly you pulled off some electrons from the plastic. Since you did the same to both, they remained with the same positive charge—proving that **like charges repel each other.**

Try to think up an experiment that will show the same results. All you have to remember is that two similar charges repel each other.

Rub one of the poly sheets with a piece of wool. The wool now picks up electrons from the plastic sheet. This illustrates that **unlike charges attract each other.**

With the same material, let's try one more experiment. Rub one sheet briskly against yourself, or against a piece of wool or fur. Place the poly against the wall of your room. It sticks and will stay there until the charge leaks away into the air. The reason is that the plastic will stick to any uncharged (neutral) body. You can do the same thing with an inflated rubber balloon.

2

Which Way Is North?

We are now going to jump 300 years and thousands of miles to China in the fourth century B.C.

It seems there was a powerful Chinese general waging war against the barbarians of the north around 376 B.C. His name was Huang-ti, meaning the "Yellow Emperor." Huang-ti was supposed to be the first to use lodestone as a compass.

There are two ways his compass could have been mounted inside his chariot. One version of the story says that a piece of highly polished lodestone rested on a piece of wood that was so polished that it was slick to the touch. The lodestone-compass turned easily on the polished mounting so it was pointing north. The other version is that the lodestone rested in a wooden bowl that floated in a tank filled with water. As the lodestone turned, it would force the bowl to turn also.

Why did the lodestone point north? This type of iron—magnetite—has permanent poles. One end always points north and the other south. Look at the drawing in Figure 3A. All of the particles are lined up in the same direction. What happens then is that the north pole of the lodestone points south. Like poles repel each other. The south pole of the lodestone points north.

If lodestone is hit hard enough, it will shift some of the particles. Some would point north, some south, and others in every other direction. The lodestone loses its magnetism (see Fig. 3B).

This first compass was used by military commanders during the Han dynasty, a ruling group that controlled China from 206 B.C. to A.D. 220. Strangely, lodestone was not used for ship navigation until 900 years later. It was employed only on land by the generals and by magicians, who used it to find the right place to erect a temple or the right burial places.

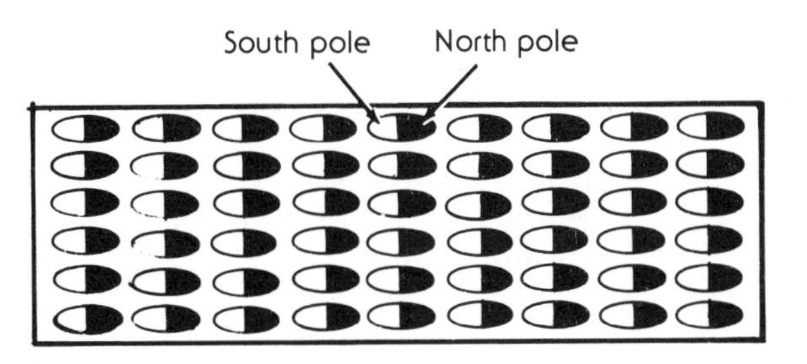

Figure 3A. Magnetized particles.

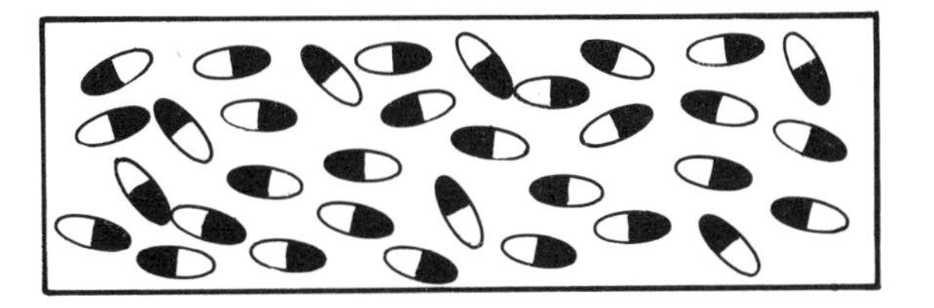

Figure 3B. Nonmagnetized particles.

Not until the thirteenth century A.D. (about the same time that Italian trader Marco Polo was exploring the Far East) was the compass employed by Chinese navigators for the first time. By this time, they had discovered that a needle could be magnetized and used in a compass by rubbing it with lodestone.

Arab sailors saw the advantages of the compass, adopted it, and brought it to Europe. This resulted in the great period of European exploration. For the first time, shipmasters could easily find their way across the sea without needing to hug the shore. Christopher Columbus undoubtedly used the compass when he left Spain trying to find an easier route to the Indies.

Experiment No. 3: MAKING A COMPASS

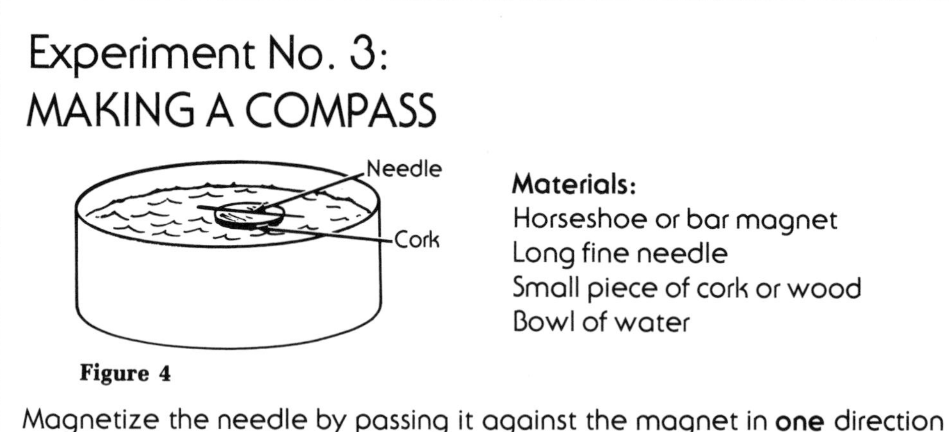

Figure 4

Materials:
Horseshoe or bar magnet
Long fine needle
Small piece of cork or wood
Bowl of water

Magnetize the needle by passing it against the magnet in **one** direction only. Then drive the needle through a piece of cork or wood just large enough so it will cause the needle to float on the water. Lower the mounted needle gently onto the surface of the water. The needle will swing at once and align itself so one end is pointing north and the other south. Bring a piece of iron or steel (a simple pot or pan will do) near your "compass." You will see the needle turn in the direction of the iron mass (see Figure 4).

In the days of Columbus, the ships were made entirely of wood, so they did not have the problem of worrying about the effect of metals on their compasses. Since today's ships are made of steel, their compasses are different from the simple compass built from Experiment No. 3.

It is very easy to make a compass like the one Columbus may have used. What we're going to do is magnetize an ordinary needle, not with lodestone but with a magnet. In the compass, the needle is delicately balanced so that it is free to move. The needle will point north unless there is a large amount of iron or steel nearby. (See Experiment No. 3 for directions on making a compass).

Experiment No. 4 will illustrate the fields of force that surround each pole of a magnet. We will see why this is important later in the story of electricity.

EXPERIMENT NO. 4: "SEEING" THE LINES OF FORCE

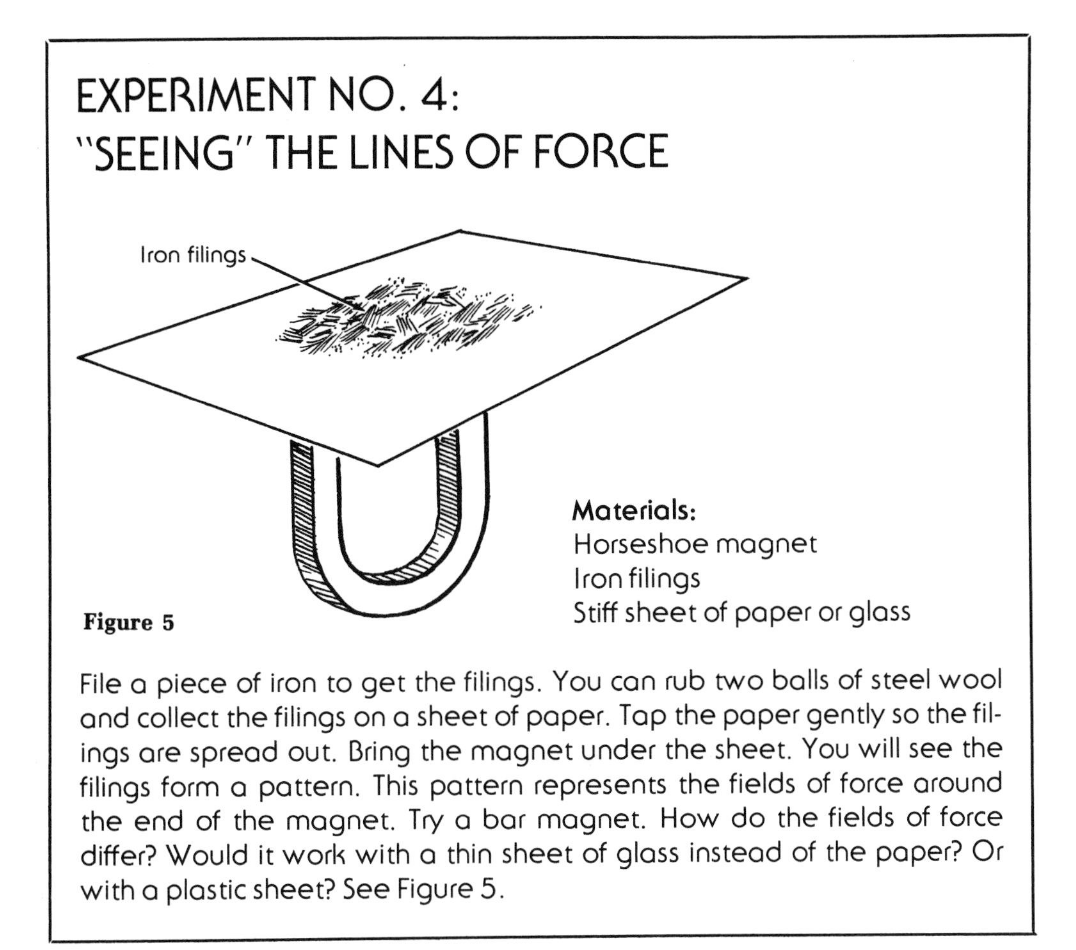

Materials:
Horseshoe magnet
Iron filings
Stiff sheet of paper or glass

Figure 5

File a piece of iron to get the filings. You can rub two balls of steel wool and collect the filings on a sheet of paper. Tap the paper gently so the filings are spread out. Bring the magnet under the sheet. You will see the filings form a pattern. This pattern represents the fields of force around the end of the magnet. Try a bar magnet. How do the fields of force differ? Would it work with a thin sheet of glass instead of the paper? Or with a plastic sheet? See Figure 5.

3

Mr. Gilbert Makes His Mark

Inventions and Discoveries

The next stop in our time travel is sixteenth-century England. While scientists had continued to study magnetic effects, no specific progress was made until William Gilbert (1544–1603) came along. He lived during the time of Shakespeare and was one of Queen Elizabeth's physicians. Like many doctors of his time, he was deeply interested in magnetism. It was thought that since magnetism had certain effects on objects, it might have healing powers for the human body.

He discovered that many substances besides amber could attract light objects, for example (listed alphabetically): amethyst, diamond, glass, jet, opal, rock crystal, sapphire, sulfur, and hard sealing wax. He found that many semiprecious stones could be added to his list. Not all of these substances attracted equally, and he carefully separated them in the order of their ability to attract.

To help gauge the ability of objects to attract, Gilbert invented the *versorium* (see Fig. 6). This is probably the first electrical instrument to be invented. *Versorium* is a Latin word meaning "turn about." That is exactly what it did: it turned around. It looked like a mariner's compass, but while the compass employed a magnetized needle, the versorium did not.

Gilbert's invention used a pointer made of almost any solid material, even light wood. Before Gilbert came up with his invention, the way to find whether a rubbed material would attract a light object was to place the two close together and see whether any motion took place. The versorium went one better. A thin pointer balanced at its midpoint would react to the presence of attraction even if there was not enough force to lift the lightest body.

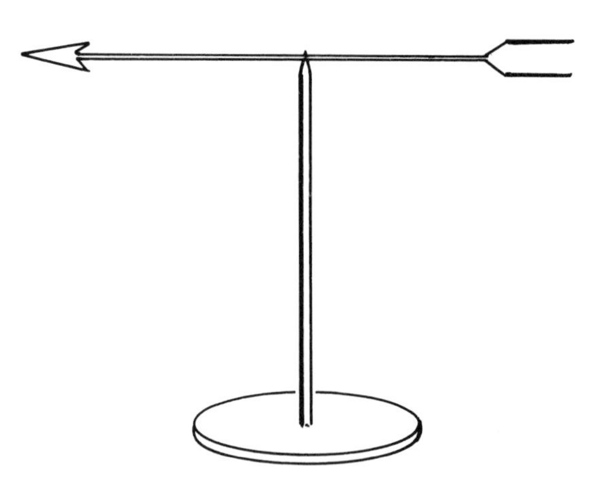

Figure 6. William Gilbert's versorium.

Gilbert describes his invention so well in his book *De Magnete (On the Magnet)*, 1600, that it is easily duplicated. (In those days important matter was always written in Latin. Educated people read and wrote in Latin as easily as we do in English.)

Experiment No. 5 shows how to build a versorium like Gilbert's.

Gilbert described his versorium in such detail that anyone could copy it. But it seems no one did. Almost 100 years were to pass before other electrics were discovered. Other writers just copied Gilbert's list without performing his experiment.

The electroscope, a modern form of Gilbert's versorium, is now used to study atomic particles.

EXPERIMENT NO. 5: GILBERT'S VERSORIUM

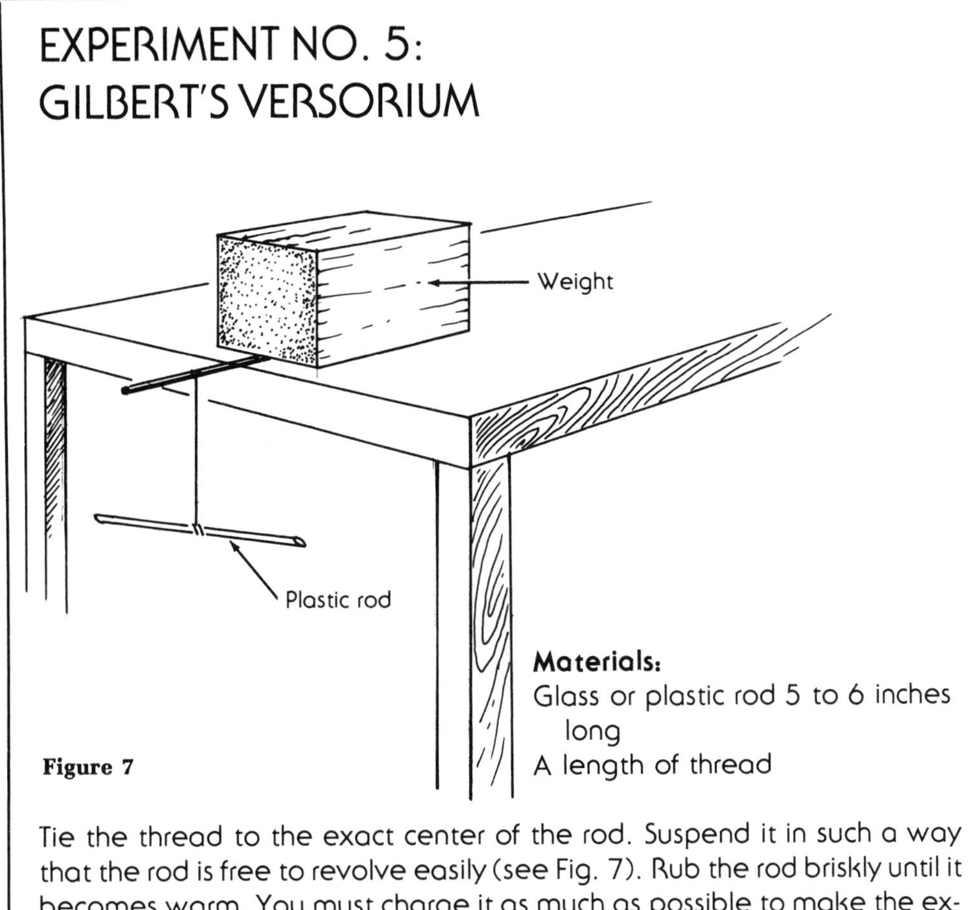

Figure 7

Materials:
Glass or plastic rod 5 to 6 inches long
A length of thread

Tie the thread to the exact center of the rod. Suspend it in such a way that the rod is free to revolve easily (see Fig. 7). Rub the rod briskly until it becomes warm. You must charge it as much as possible to make the experiment work successfully. Wait until the rod is hanging motionless. Bring a piece of paper close to it—but keep them slightly apart. The rod will turn toward the paper and the paper will lift toward the pointer. Try other objects as a substitute for the paper.

Carry the experiment a step further. Take an empty plastic ball-point pen. Rub it against a wool shirt or rub it with a piece of fur. Then bring it close to the pointer. It will revolve away as fast as it can. You can get the pointer to turn many times by chasing it with the pen. This is another proof that two **like** charges repel each other.

Apart from this invention, Gilbert deserves credit for making up the word *electricity.* He used the word to describe objects placed near his versorium. Those that were not attracted he called *nonelectrics.* Those that were attracted, such as paper, straw, and sealing wax, were called *electrics* because they were attracted to the pointer and made it move.

He discovered other important things during his studies. He found that it was not the heat given off by the rubbing that made amber attract other light objects, but the friction. He also dispelled the theory that air was displaced by the amber or that the attractive quality was owned only by the amber. He proved that substances very different from amber were also electrics.

A Scientific Approach

Gilbert was different from other scientists of his time in his attitude toward his work. After he developed a theory, he would perform various experiments himself in order to find out whether the theory was right or wrong. Most other scientists (or "philosophers," as they called themselves) would work out a theory, but felt it was beneath them to become "workmen" and build the necessary equipment to prove their point.

Gilbert also developed the idea that extensive experimenting should come before coming up with a basis for a theory. You might *think* you know the reason for the behavior of a substance or the explanation for the action of some natural force. But in order to prove that your theory is true, you have to make up an experiment and repeat it many times, changing it slightly each time.

As an example: Let's say you believe that it is the movement of the trees that makes the wind. You say, "Look, the trees move and there's a wind. I can feel it. Doesn't it prove I'm right?"

To prove or disprove your theory, you must do an experiment. You go to places where there are no trees. There you can feel the wind on your face. Since there is a wind and no trees, it must prove that the wind exists without trees. Of course, to be really scientific, you would have to go to several different places, such as plains, deserts, and mountaintops. You repeat the experiment to check on the accuracy of your observations.

And that's what William Gilbert did. He brought a truly scientific approach to the study of what in those days was still called magnetism, but which today we know to be static electricity. In his honor, a unit of magnetic intensity is called a *gilbert*.

4

The Electrifying Seventeenth Century—I

Electric Generators

Otto Von Guericke (1602–1686), a German experimenter, constructed the first electric generating machine (see Fig. 8). In 1660 he not only generated electricity in greater amounts than ever before by means of his machine, but he also demonstrated that it could be *transmitted.*

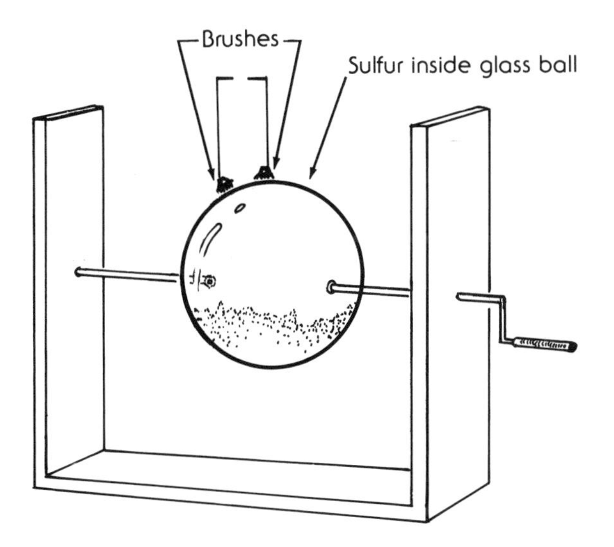

Figure 8. Von Guericke's machine.

The machine consisted of a large ball of sulfur inside a glass globe. Mounted on a shaft, the glass globe was rotated at high speeds by means of a crank. It was excited by friction by applying a cloth pad. Two brushes touching the globe allowed sparks to leap across a narrow air gap between two metal balls.

By moistening a string, Von Guericke was able to transmit electricity several feet over the string. Wetting it made it a better conductor of electricity.

Instead of using a sulfur ball inside a glass globe, the experiment can be duplicated today by using a large plastic disk. The center of the disk is drilled and a shaft inserted through the hole. A crank turns the disk at high speed.

Some schools have a Wimshurst machine that operates on the same principles as Von Guericke's own invention. If your school has one, you may wish to ask a teacher to show you how it works—for example, the machine can, quite literally, make your hair stand on end! (The Wimshurst machine is one of those that should be examined under knowledgeable supervision; improperly used, it could deliver a substantial electric shock!)

How Does "Magnetism" Work?

William Gilbert believed that the earth was a huge rotating magnet. However, many learned people disagreed with him. They felt that our planet was the center of the entire universe, and that the sun and all the other planets revolved about us. The earth was immobile, they said.

One man who claimed the earth was stationary was Italian experimenter Niccolo Cabeo. In 1629 Cabeo wrote a treatise in which he stated that the earth was immobile. Since he felt that Gilbert was wrong in believing the earth rotated, he tried to find more mistakes in Gilbert's work. Ironically, while trying to find these errors, Cabeo enlarged the list of electrics from Gilbert's own list. He also found that objects attracted to a rubbed electric would not only fall off after a few moments, but would then be repelled from several inches away. This effect had not been noticed by Gilbert. We now know that after the numbers of electrons in two objects have become equalized, they will repel each other.

Cabeo tried to explain how "magnetism" worked, since there was no real theory yet on why objects sometimes clung to each other and sometimes did not. Cabeo's idea was that when an electric was rubbed until it became hot, it created a force that drove away the air around it. This would allow two objects to come together, since there was no air to stop their joining.

How could he prove his theory?

At about this time the barometer had been invented. It was used to measure atmospheric pressure. We still employ it to help us forecast the weather, as the atmosphere changes in pressure when a storm approaches. To make a barometer work, a tube holding mercury had to be emptied of air so the mercury could respond easily to air pressure, creating a state of near-vacuum. This was an ideal way to find if "magnetism" operated when no air was present.

Experiments were carried out, but so much air leaked into the tube holding the mercury that no one could tell if Cabeo was right or not. Otto Von Guericke, whom we just met, invented a pump to keep the air out of a bag mounted on top of the column. The bag allowed a hand to reach in to rub the amber used in the experiments. But the pump was not efficient enough.

It was not until 1675 that Robert Boyle (1627–1691), an Irish physician, developed a greatly improved pump. He found that "magnetism" works just as well in a vacuum as it does in air.

Hauksbee's Glowing Discoveries

Boyle's pump created a great deal of interest in the effect that air, or the lack of it, had upon "magnetism." One of the men who became interested in Boyle's work was Francis Hauksbee. While we don't know much about him, we do know that he published a book in 1709 containing a variety of experiments. From his papers, it seemed Hauksbee had little formal schooling—but he more than made up for that lack by his ability to build instruments and his unusual skill in developing experiments.

During that period in England, many were experimenting with what was called "barometric light." A barometric tube was partially filled with mercury. Air was pumped out of the top of the tube. Experimenters found that shaking the tube vigorously produced a glow or a flash of light. Various theories were introduced to account for this phenomenon. We know today that it is electrical in nature and resembles the light we see in a neon sign. We also know that the light in the tube appeared because the tube was electrified by the action of the mercury rubbing the glass as it was shaken.

Hauksbee experimented endlessly using equipment he invented and built for that purpose. As a result of this work, he came to the conclusion that it was the action of the mercury that created the light. He noticed that the light occurred only when the droplets of mercury slid down the sides of the tube, not when they were motionless.

Not satisfied with these discoveries, he varied the amount of air he left in the tube. He found at a certain pressure—1/2 to 1/20 of an *atmosphere* (the amount of air pressure at sea level)—he could obtain the most light. The effect he obtained was similar to St. Elmo's fire, which appears on masts of ships before an electrical storm. But as yet, no one connected those very weak flashes of light with a phenomenon such as lightning.

In a novel experiment, Hauksbee removed the air from a glass globe and caused it to rotate while rubbing it with a cloth—probably wool. The light that resulted was so bright that he could read large letters in a dark room.

He found that the hotter the glass became, the greater was the distance from which light objects—paper, feathers, straw—would be attracted to it. Bringing his face close to the electric, he felt something like a breeze across his cheeks. This is known today as an "electric wind." It consists of charged air particles that are repelled by the rubbed electric.

All through the centuries during which "magnetism" was studied, the electric was rubbed by hand. Hauksbee produced more electricity than was ever possible before by attaching a large wheel by means of a belt to a shaft on which was mounted a glass globe. The large wheel was cranked by hand. Each revolution of this wheel made the globe rotate at a speed he estimated to be 29 feet per second. This speed caused the globe to become very hot. The heat produced much more electricity than could be gotten by simply rubbing by hand.

Hauksbee's work is of great importance because many of his discoveries challenged the beliefs of his time about electricity. He raised many questions he could not answer. This opened up a new field of inquiry. The first to attempt to answer Hauksbee's questions was Stephen Gray.

5

The Electrifying Seventeenth Century—II

The Excited Electrics of Stephen Gray

Not a great deal is known about Stephen Gray, another contributor to the study of electricity. We know he was born sometime between 1666 and 1695, and died in 1736. Otherwise, all we know about him is derived from some unpublished letters and communications to the Royal Society of England. In his first published paper, which appeared in 1720, he listed some new electrics, including silk, wood, and hair—both human and animal. He also described the following experiment.

Take a hollow glass rod about two feet long. Rub it very briskly with a piece of fur or silk in a dark room. It works best of all, like most experiments in static electricity, on a cold, dry day. You will find that the light coming from the tube will follow your hand if you bring it close enough, and a spark will pass from the tube to the finger held close to it.

Gray showed that you could electrify something entirely new—stiff paper. It had to be heated as hot as possible and then rubbed. It would then become electrified.

About ten years later, he wrote a paper about another discovery of his. He rubbed a glass tube, then brought other objects to it, and found that the objects that came in contact with the tube had the same "magnetic" properties as the rubbed tube itself.

Another experiment showed that an excited electric was capable of "emitting electricity to another body but without the accompanying light." The tube he used for that purpose was 3½ feet long and about an inch in diameter. He corked the tube at each end to find out if there was any difference in the amount of electricity if the tube was stopped up or open. He found there was no difference.

Holding a feather near the upper end of the tube, he found that the feather would be attracted to the cork and be repelled just as it was by the tube itself. From this he concluded that what he called "an attractive virtue" was communicated to the cork by the rubber tube. Today we don't call this attractive virtue, but *electric conduction.*

Attractive Virtues and Electric Conduction

Gray then realized that the ability to attract had been communicated to the cork because it was touching the tube. He formulated the theory that an object that touches an electrified body will itself become electrified. This idea led to a whole new group of experiments. What he wanted to find out was whether this ability to attract could be transferred not only to bodies touching each other but even through long conductors in between.

While the idea of *conduction*—of electricity being transmitted from one point to another through a wire or other medium—may sound very simple to us, it was a very new idea in Gray's time. It was also very difficult to prove, since they were dealing with very small, fluctuating currents of static electricity rather than the type of sustained current that we can now get even from the smallest battery.

In addition, they had no efficient way of measuring a current of electricity. For example, we have meters today that are capable of measuring even small amounts of electricity. So the combination of lack of delicate instruments and irregular and low voltages made it difficult to carry out experiments that would be extremely simple for you and me.

Now let's return to Mr. Gray's problem of trying to prove his theory. He took an ivory ball about an inch in diameter, drilled a hole into it, and thrust a stick into the hole. He pushed the other end of the stick into the cork capping the "electrified" tube. He rubbed the tube. The ball attracted and repelled the feather even better than the plain cork. He used longer and longer sticks. He used an 18-foot fishing pole. It was just as good in conducting electricity. Wire was very rare in those days but he managed to get some, and found the feather was attracted to whatever part of the wire it came near.

In further experiments, Gray suspended a ball from the tube using cotton thread about three feet long. The thread conducted electricity and the ball attracted and repelled light objects. Whether he used chalk, lodestone, or a brick, he found it had "attractive virtue."

You will remember that Gilbert attempted to discover whether or not different metals were electrics by rubbing them while he held them in his hand. Many had not shown any "attractive virtue," so he concluded that they were nonelectrics. But Gray found that they were indeed capable of being called electrics because he did the experiment without actually touching them. He was able to show that the body was a conductor and the "attractive virtue" of the metals was conducted away by their being held in the hand.

To get longer transmission length, he decided to lay out his line horizontally. He went so far as to place an ivory ball at the end of a thread 765 feet long (over one tenth of a mile). Again the ball at the end showed the same ability to attract and then repel a feather, or even pull up thin pieces of brass ("shims"). He also found that by wetting the thread, he could get more "magnetic" strength and that the attraction of the brass to the ball worked even when the ball was *not touching* the brass. When the metal was not touching, the attraction was only temporary. When there was a connection—that is, when the metal was touching the ball—the attraction lasted much longer.

However, when Gray tied the line to the beams of the room in which he was conducting his experiments, no matter how hard he rubbed his glass tube, he lost the attraction. From this he concluded correctly that the beams were conducting the electricity away.

In a later experiment, Gray took an iron rod sharpened at both ends and suspended it from silk threads. When he brought an electrified tube near the bar, he saw cones of light shining from the bar in his darkened room. He made note of the crackling noises that accompanied the spark. The noise and the spark reminded him of the roar of thunder and lightning bolts.

That observation marked the beginning of the realization that lightning was similar to the static electricity being studied. But many years would pass before it would be shown that these were merely forms of the same phenomenon—electricity.

Now let's jump on our magic travel machine and be transported to Paris in the year 1734. The scene is the French Academy, where men of science met to discuss their views and the results of experiments in all the known sciences of that period.

Everything Can Be Electrified

Charles Du Fay (or Dufay, as it is sometimes spelled) was born about the same time as Stephen Gray, between 1666 and 1698. He died just three years after Gray, in 1739.

He was a dedicated experimenter in the electrical sciences. He was also interested in chemistry, botany, physics, and anatomy—in other words, he was curious about almost every science then known. He was a member of the Royal Society in London, but he was also a member of the French Academy of Sciences.

It was in Paris in 1734 that he read a paper that dealt with the findings of William Gilbert and others who said that every substance was either an electric or a nonelectric. About a year before, Du Fay had heard of Gray's experiments and decided to try some of them himself. By 1734 his experiments had convinced him that *all* substances could be electrified if they were heated first and then rubbed with a cloth or fur. He said that even a conductor could be electrified if it was insulated.

For example, take a bare wire and electrify it by connecting it to a rubbed glass rod as we described before. Now lay that wire on the ground. There will be no voltage at the end of the wire. Why? Because the electricity has leaked away into the ground. Insulate the wire by encasing it in rubber or silk, and the voltage will be just as great at the end of the wire as it was at the tube.

Gray had found that water could be electrified. Du Fay eventually showed that other liquids could be made to become electrics. He also proved that a wet string could conduct electricity better than a dry string. He proved it by electrifying a cork ball that was connected to the glass tube by a string 1,256 feet long. That's almost a quarter of a mile of wet string!

One of Du Fay's most exciting discoveries was that glass and silk were insulators and could be employed to support a conducting line. Today we use glass or ceramic insulators on the poles that carry high tension wires. Many lamp cords are covered with silk for the same reason—to keep the electricity from "sneaking" away. Rubber and plastic are most often used as insulation for wires.

The world in general was becoming familiar with the wonders of electricity. However, no one really knew what it was. It could be explained by thinking of it as a force that surrounded an object, or as a fluid—in fact it was often called the "electric fluid."

Du Fay's most important discovery was yet to come. He used a piece of gold leaf—a sheet of gold so thin that, when applied to statues, it makes them seem to be made of solid gold, yet every line on the statue can still be seen. When Du Fay brought the gold leaf against his electrified glass tube, the gold sprang away and floated above it in the air.

His next step was to make a ball of rubbed copal. Copal is a solidified resin that comes from various tropical trees. It is similar to amber in that they both are resins. He brought the copal against the floating gold leaf. Since they were both electrified, he expected them to be attracted to each other. To his surprise, they were not. From this he got the idea that there must be two kinds of electricity: one was *vitreous* (coming from glass), and the other was *resinous* (coming from a resin). His conclusion was that a substance electrified with resinous or vitreous electricity would repel substances with the same type of electricity. The contrary, he said, was also true. Substances would be attracted to each other if one had resinous electricity and the other had vitreous electricity.

Du Fay said that vitreous electricity was produced by rubbing glass, diamonds and other precious stones, wool, and the hair of animals. By rubbing resins such as amber, copal, thread, and paper, you obtained resinous electricity.

He came up with a conclusion that we know is absolutely correct: Everything and everybody contains electricity. What he did *not* include was lightning—a strange oversight!

Performing Experiment No. 6 will give you a good idea of Du Fay's own approach to studying electrics.

EXPERIMENT NO. 6: STUDIES IN ELECTRICAL CONDUCTION

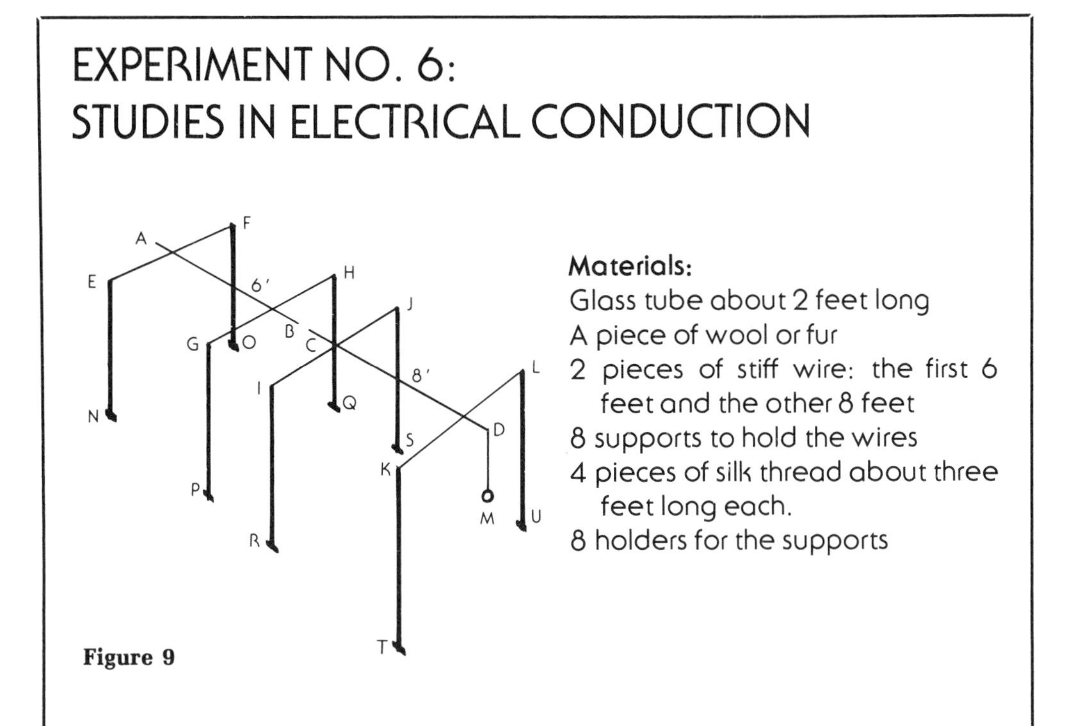

Figure 9

Materials:
Glass tube about 2 feet long
A piece of wool or fur
2 pieces of stiff wire: the first 6 feet and the other 8 feet
8 supports to hold the wires
4 pieces of silk thread about three feet long each.
8 holders for the supports

See the sketch in Figure 9 on how to set up the experiment. **A–B** and **C–D** are made up of two pieces of very heavy-gauge wire. It doesn't matter whether they are insulated or not. **E–F, G–H, I–J, and K–L** are silk threads. Their length is unimportant. The threads are tied to the supports **N, O, P, Q, R, S, T,** and **U.** Their height is also unimportant. Their function is to keep the heavy wires off the ground. Point **A** is connected to a static generator, such as the electrophorus described in Experiment No. 8. **M** is a cork ball suspended by a silk thread. Notice that Points **B** and **C** are close together but **do not touch.** Once your static machine is loaded, you discharge it into Point **A.** Then bring a feather or other light objects up to the cork ball. Du Fay found that an electrified feather is first attracted to the ball and then repelled by it.

As you may have already guessed, this experiment shows not only that electricity can be conducted between two points, but that it can jump a gap between conductors—that is, the two wires. While the electricity conducted in Du Fay's experiment was not strong enough to do any actual work, the principle that it proved was to become extremely important later on.

6

Jars and Kites to Frogs and Batteries

During the eighteenth century, electricity was of interest to everyone. They talked about it much as we do today about home computers. No one thought electricity could do anything useful, but it was a source of wonder nevertheless. People flocked to the lectures given by so-called "professors"—since there was no radio, no movies, and no TV, lectures of all kinds were well attended. This was a way for people to keep up with the world around them.

To get more amazing results—bigger sparks on the darkened stages and louder crackling sounds—the static electric generators were being made more powerful.

Most of these machines consisted of a glass ball mounted on a shaft. The ball was made to revolve by means of a crank—very much like Hauksbee's generator.

A collector was invented later that employed a comblike affair. A number of thin metallic points touched the revolving ball.

Some of these machines could generate tremendous sparks. One experimenter touched his machine one day and it threw him across the room! Unsurprisingly, he said he would never repeat the experiment.

One favorite demonstration during a lecture was to have several people holding hands. One would touch the machine while it was cranked up. The last person on the human chain would touch someone in the audience. You can imagine what happened!

Severe Shocks from Jars

To meet our next experimenter-scientist, we must travel to Germany in the year 1745. The man's name is Ewald von Kleist. People believed then that an electrically charged object lost its charge in the open air. This, they thought, was due to evaporation of the "electric fluid." Von Kleist, however, wondered: If a collector was completely sealed and electrified, would it keep its charge longer?

He began by half-filling a glass bottle with water and sealing it with a cork. A long nail was inserted through the cork into the bottle so that the nail touched the water.

Von Kleist then held the bottle in his hand, bringing it close to a static generator so the nail made contact with the generator and conducted electricity between it and the water. He pulled the bottle away and then let the nail *almost* touch a nonelectrified object. A great spark leaped through the air from the nail to the object. He touched the nail with his other hand and received a severe shock.

Substituting liquids such as mercury and alcohol for the water, he got bigger sparks. He wrote later, "I touched the nail and received shocks so severe that my shoulders and arms were stunned."

What was most interesting was that if no object touched the nail, the contents of the bottle remained charged for many hours. This was the first time that static electricity was stored.

Just a few years later, a famous teacher of mathematics, Pieter Van Musschenbroek (1692–1761), described an experiment that was

EXPERIMENT NO. 7: CONSTRUCTING A LEYDEN JAR

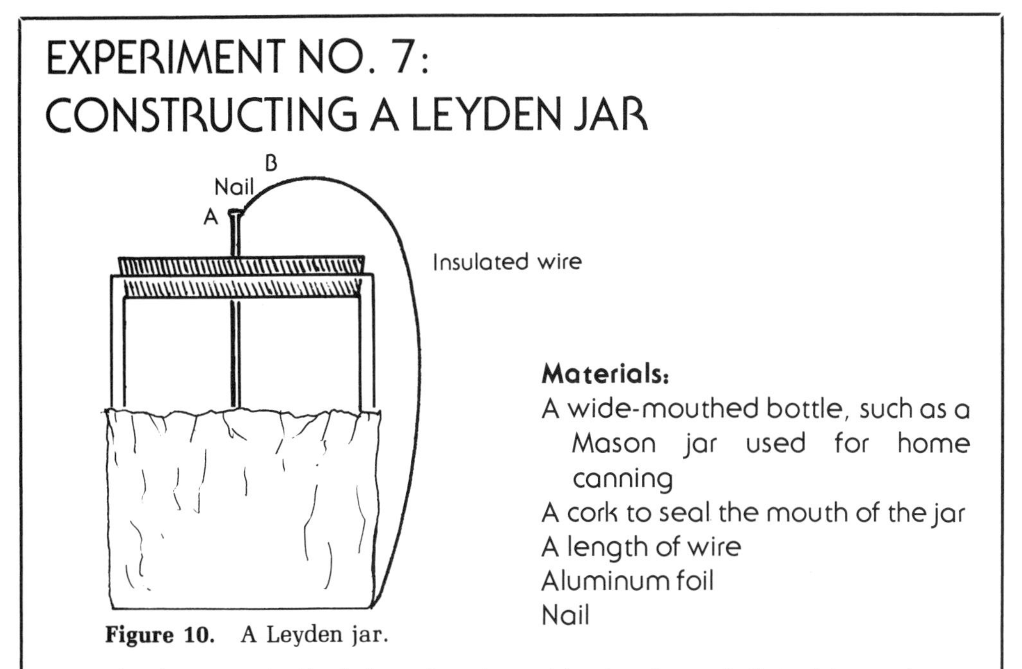

Figure 10. A Leyden jar.

Materials:

A wide-mouthed bottle, such as a Mason jar used for home canning
A cork to seal the mouth of the jar
A length of wire
Aluminum foil
Nail

Line the bottom half of the glass jar with aluminum foil, inside and out. Press it tightly against the glass. Seal the jar with a large cork or with paraffin wax (used for home canning). Insert a nail or a metal prod long enough to touch the bottom of the jar and the foil. It should emerge about an inch or two above the seal. Take a wire with about one inch of insulation removed from one end and place it between the glass and the outside foil. Make certain there is a good solid contact between the wire and the aluminum foil.

The simplest way to charge a Leyden jar is to rub a glass rod with a piece of fur, silk, or nylon. Discharge it at Point **A.** Do this over and over until you feel you have stored a good amount of static electricity. Do not touch Point **A** with any part of your body or you will discharge the jar. If you do touch the point of the jar, you will feel a slight tingle just as you do when you touch a metal object after walking across a rug on a cold, dry day.

Charge the jar only enough to see a small spark jump across the gap. This unit is not designed to produce a large amount of static electricity, only to illustrate the principles.

Holding the outside wire where it is insulated, bring it close to Point **A.** If the jar has been charged enough, a tiny spark will jump from **A** to **B.** On a cold, dry day you may also hear a tiny crackle.

To do further experiments, you will have to recharge the jar all over again.

The jar can also be charged several times by means of the electrophorus or static machine described in Experiment No. 8 (Fig. 11). Lay the jar on an insulated surface—a few sheets of dry newspaper will do fine.

You now have a source of static electricity. With the jar, you should be able to invent many interesting experiments of your own. Have fun!

Warning: Do not put a meter between Points **A** and **B.** The sudden surge could burn the meter out.

EXPERIMENT NO. 8: YOUR OWN CHARGER: THE ELECTROPHORUS

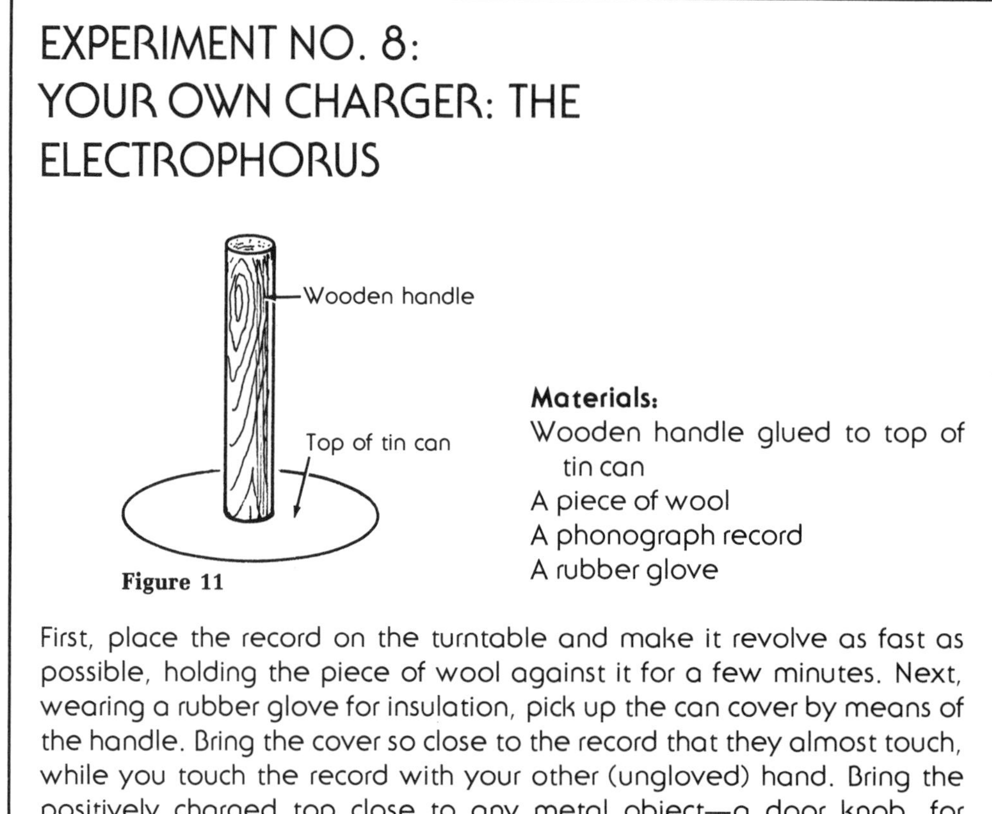

Figure 11

Materials:
Wooden handle glued to top of tin can
A piece of wool
A phonograph record
A rubber glove

First, place the record on the turntable and make it revolve as fast as possible, holding the piece of wool against it for a few minutes. Next, wearing a rubber glove for insulation, pick up the can cover by means of the handle. Bring the cover so close to the record that they almost touch, while you touch the record with your other (ungloved) hand. Bring the positively charged top close to any metal object—a door knob, for example. A spark from the top will jump to the metal object. The stronger the charge on the can cover, the bigger the spark. It's best done on a dry, cold day.

the same as von Kleist's. Because Musschenbroek was better known at the time, he was credited with the invention of the storage jar. It was called a Leyden jar, after the University of Leyden, in Holland, where Masschenbroek taught.

The Leyden jar is the forerunner of what we now call a *capacitor*. Without it, very few electronic circuits would work. For example, a capacitor is used to store electricity in the unit that operates the flash for a camera. Two or more tiny batteries pour a small amount of electricity into the capacitor. When it is fully charged, it is capable of generating a great burst of light for a fraction of a second. Radios, TV sets, and computers all depend on the capacitor to hold electricity for hours or even days.

Experiment No. 7 shows how to make your own Leyden jar (Fig. 10). While it won't store a great deal of electricity, it will produce just enough to make some interesting experiments.

The Leyden jar showed for the first time that electricity could be stored for later use. In addition, a much larger quantity could be built up in a jar than could be gotten by the one-time use of a friction machine. Successive charges could be conducted into the jar until it could handle no more, the same way you would fill a glass with water one drop at a time.

The Curious Ben Franklin

Sometime around 1746, a Professor Spencer of Edinburgh brought his lecture on electricity to the American colonies. At a lecture in Philadelphia, one of the most interested spectators was a man who has been called the first American scientist. Benjamin Franklin

(1706–1790) was curious about everything. His mind absorbed knowledge, and the results of his thinking were many and varied. He started the first magazine in this country. He invented the stove that bears his name. He began the first circulating library and the first scientific society. He was also American ambassador to France.

Still he found time to attend Professor Spencer's lecture. He was so caught up by the marvels shown on that stage that he ordered a static generator of his own. He also obtained a Leyden jar and conducted his own experiments.

Franklin repeated many of Du Fay's experiments. He did not know that Du Fay had declared there were two kinds of electricity: vitreous and resinous. As a result of his own experiments, Franklin concluded that there were indeed two kinds of electricity. He called them *positive* or plus (+) and *negative* or minus (–).

He said that electricity was not created by rubbing the glass tube of the generator. It was merely transferred. He went on to state that when an unelectrified object was rubbed, it did one of two things: It either gained electricity and reached a positive state, or it lost some of the "electric fluid," leaving the object in a negative state. This idea that electricity could be created and/or destroyed was an important one.

In one experiment, Ben Franklin suspended a cork ball from a silk thread. It hung between a wire attached to the inner coating of the Leyden jar and a wire attached to the outer coating. The ball swung first to one wire and then swung to the other. It kept up this pendulum effect until the Leyden jar was de-electrified. This proved the existence of the two types of electricity: The inner coating was negative and the outer was positive.

Benjamin Franklin was a serious experimenter and made many worthwhile discoveries, but he is best known for his kite experiment. It was in 1752 that he flew the famous kite. He picked a day when a storm was about to break. At the top of the kite, he fastened a stiff wire pointing up. At the other end of the string, he tied a metallic key so it hung close to a Leyden jar. It started to rain. The moistened string began to conduct electricity. It was very fortunate for Franklin that there was no lightning. The storm ahead of the storm gave him enough electricity to prove without a doubt that electricity and lightning were the same. Sparks jumped from the key to the jar until the jar was filled with "electric fluid."

Not as fortunate as Franklin was a Russian scientist of that period, who was killed by lightning when he held a rod high in the air during a thunderstorm.

(For hundreds of years it had been believed that evil spirits rode the storm and created lightning. When a severe lightning storm approached, a curious custom took place. Bell ringers were summoned to the churches to ring the bells. What often happened was that the bell ringers were killed by the lightning. The bolts of lightning struck the church spires. The bells were made of metal, and the lightning traveled through the bells down the wet rope in the hands of the ringers. So many ringers were killed that laws were passed to stop this deadly custom.)

The Wise Mr. Franklin

Another of Franklin's experiments proved that a pointed rod is a better conductor than a ball. You can demonstrate this very easily. Walk across a rug shuffling your feet. Touch a doorknob with the tip of your finger. You will feel a shock. Repeat your shuffle, but this time touch the doorknob with your fist. You will get little or no shock.

A similar experiment led Franklin to believe that a long pointed rod suspended above a house or a barn would attract lightning. The rod was connected to the ground by means of a wire. The lightning would go down the lightning rod, through the wire, and be absorbed safely into the ground.

Being a good businessman, Franklin went immediately into the manufacture of lightning rods. He sold them throughout the colonies.

There is no doubt that Franklin's lightning rod kept many houses and barns from burning down. But lightning does not always behave the way it is expected to. Sometimes houses with lightning rods were struck, and no one knew why. We understand now that lightning may come through an open window or even follow a current of warm air.

His was such an inventive mind that he was not satisfied with what he had accomplished, but went on to invent an electrostatic motor (see Fig. 12).

gypsies—left the glass and moved to the silk. That left the glass positively charged. The silk had extra electrons so it was negatively charged.

But does the same thing happen when you rub amber, wax, or rubber with wool? No. The result is a negative charge in the rubbed substances.

Connect two pieces of metal with different charges by means of a wire. The electrons will

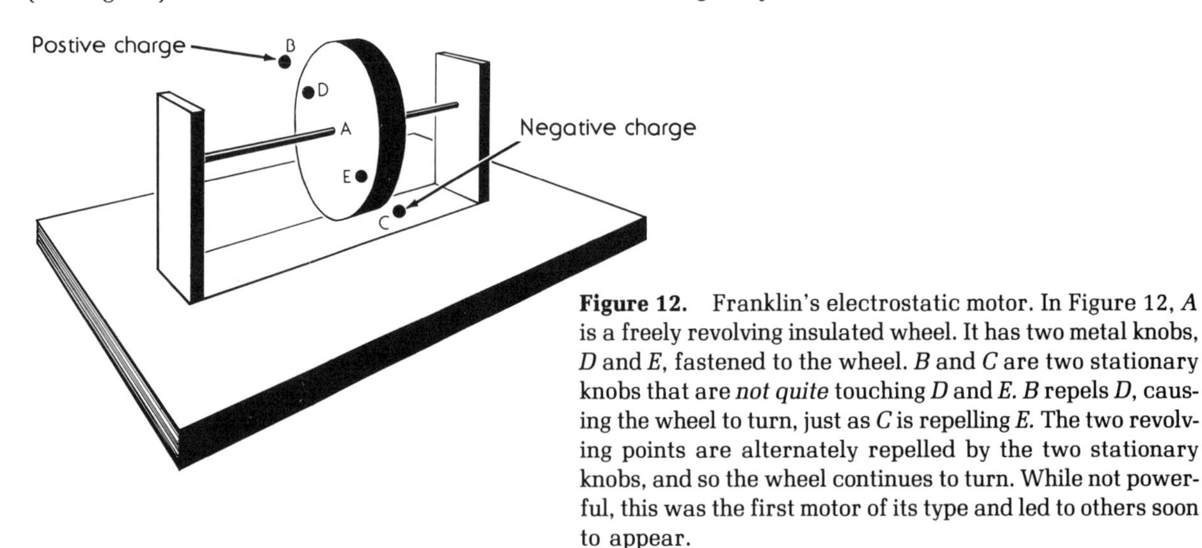

Figure 12. Franklin's electrostatic motor. In Figure 12, *A* is a freely revolving insulated wheel. It has two metal knobs, *D* and *E*, fastened to the wheel. *B* and *C* are two stationary knobs that are *not quite* touching *D* and *E*. *B* repels *D*, causing the wheel to turn, just as *C* is repelling *E*. The two revolving points are alternately repelled by the two stationary knobs, and so the wheel continues to turn. While not powerful, this was the first motor of its type and led to others soon to appear.

Franklin's theories about positive and negative electricity are still used today. All electricity is made up of two different particles in equal amounts. Some are *protons*, with a *positive* charge. Others are *electrons*, with a *negative* charge. You can think of the protons in a solid substance as being fixed in place, while the electrons are free to wander.

Remember the friction experiments described earlier, in which glass was rubbed with silk? The electrons—those wandering

flow along the wire like a fast moving stream. They move from the negatively charged metal to the positive. This produces an electric current.

Later we will discuss an experiment you can do that will show current flow when you join two different metals together.

But we have Luigi Galvani waiting for us in Italy, ready to show us his experiments. So we can't delay any longer.

Galvani's Jumping Frog

Like many men of his time, Luigi Galvani (1737–1798) was involved with several sciences. He was professor of anatomy at the University of Bologna, but he was also familiar with chemistry and physics.

There are several versions of how he made the discovery that bears his name, Galvanic electricity. The stories agree only upon the date, 1780. According to the most probable version, he was conducting a class in anatomy in which he was dissecting a frog. It seems he put the frog down near an electrical machine that had been used for some previous experiments. His wife, who was the daughter of Galvani's own teacher, was in his class. She noticed a spark pass between the frog and the electrical machine. When a scalpel touched the frog's nerve center, the leg twitched.

Galvani was never able to explain how it happened or what caused it. He was so interested in this phenomenom that he tried many other experiments. He substituted other metals for the steel scalpel. The frog's muscles continued to jump. He became convinced that it was electric in origin, but thought incorrectly that the nerves of the frog contained the electricity.

Eleven years later, he published the results of his experiments. By that time, many had heard of his work. One of those interested in Galvani's frog was another Italian, Alessandro Volta.

Volta's Battery

A professor of physics at the University of Padua, Volta (1745-1827) had a different idea as to where the electricity came from in Galvani's experiments. He was certain that it did not come from the frog but from the metals—the steel knife or the metal table on which the frog was laid. This began a big argument between the two men. Each had followers who claimed the other side was ignorant or crazy. We now know that there was some truth in both Volta's and Galvani's ideas.

Volta is better known as the inventor of the voltaic pile. We call it an electric cell, but it is more commonly known as a battery.

It was on March 20, 1800, that Alessandro Volta sent a letter to the Royal Society of London describing his discovery. He had made a stack of a zinc disk and a copper disk, with a paper or leather disk in between. The in-between disk had been soaked in a salt solution or a mild acid such as vinegar or lemon juice. He built up a high stack alternating the zinc, paper, and copper disks.

Do you know why he started with zinc and ended with copper? You guessed it! There had to be two different metals to form the ends of the voltaic pile.

You can duplicate what he did by following Experiment No. 9 (see Fig. 13).

Volta and Galvani Are Right on Electricity

Before Volta invented the chemical cell, experiments were generally done with static electricity. Here, for the first time, was a way to work with dynamic electricity and to have it any time it was needed, rather than having to crank a static generator to fill a Leyden jar. And there's something else that's inconvenient about static electricity as compared to dynamic electricity: Static electricity comes in large bursts. It cannot be controlled. Using static electricity is like emptying a tank of water over your head to get a drink—a little impractical.

Volta and Galvani were both right, because the moisture in the frog's flesh was the acid in the experiment. Volta did not know that by using two dissimilar metals separated by a conducting liquid, the electrons escape through the liquid. The metal that is less acted upon is charged to the higher voltage potential. Nevertheless, Volta's contribution is an important one to the science of electricity, even if he did not know why it worked the way it did.

His discovery is the forerunner of the *primary cell*, which delivers current as the result of an electrochemical reaction. This reaction is not efficiently reversible. This means that it cannot be easily recharged. On the other hand, a *secondary cell*, like a car battery, can be charged, discharged and recharged thousands of times.

The *volt*, the unit of electrical force, is named after Alessandro Volta.

At one time, an electric current was called a *galvanic current* in honor of Luigi Galvani. So the instrument made to measure or indicate the presence of current in a circuit was called a *galvanometer*. It is now called an *ammeter*, and the unit of current is the *ampere*, or *amp*, for short. If it measures small amounts of current, it is called a *milliammeter*. *Milli-* means "one thousandth," so a milliammeter measures thousandths of an ampere.

EXPERIMENT NO. 9: YOUR OWN BATTERY

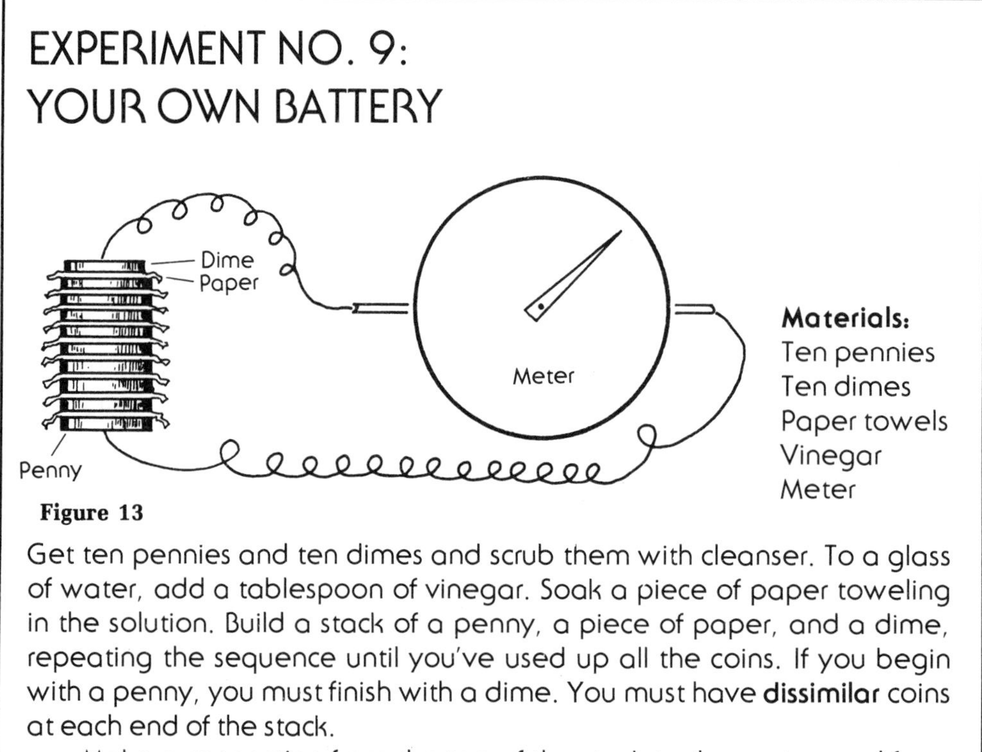

Figure 13

Get ten pennies and ten dimes and scrub them with cleanser. To a glass of water, add a tablespoon of vinegar. Soak a piece of paper toweling in the solution. Build a stack of a penny, a piece of paper, and a dime, repeating the sequence until you've used up all the coins. If you begin with a penny, you must finish with a dime. You must have **dissimilar** coins at each end of the stack.

Make a connection from the top of the stack to the meter, and from the bottom to the other side of the meter.

You should get a momentary indication on the meter. It will make the needle move only a fraction.

Double the size of the stack. Do you expect to obtain twice as much current? Why the vinegar?

A Lemon of a Power Source

A recent TV commerical showed a man demonstrating what he claimed was a "new" source of electric power. He said that he extracted electricity from a lemon. From what you just found out about electricity, you are aware that his claim to a "new" source is silly. There is nothing new about it.

Of course you can extract electricity from a lemon. Why not? Lemon contains citric acid, just as does vinegar or sour milk. So when you place two wires made of *different* metals into the lemon, you will obtain a tiny current. Experiment No. 10 shows how this works, and Figure 14 shows how it looks.

EXPERIMENT NO. 10: THE ELECTRIC LEMON

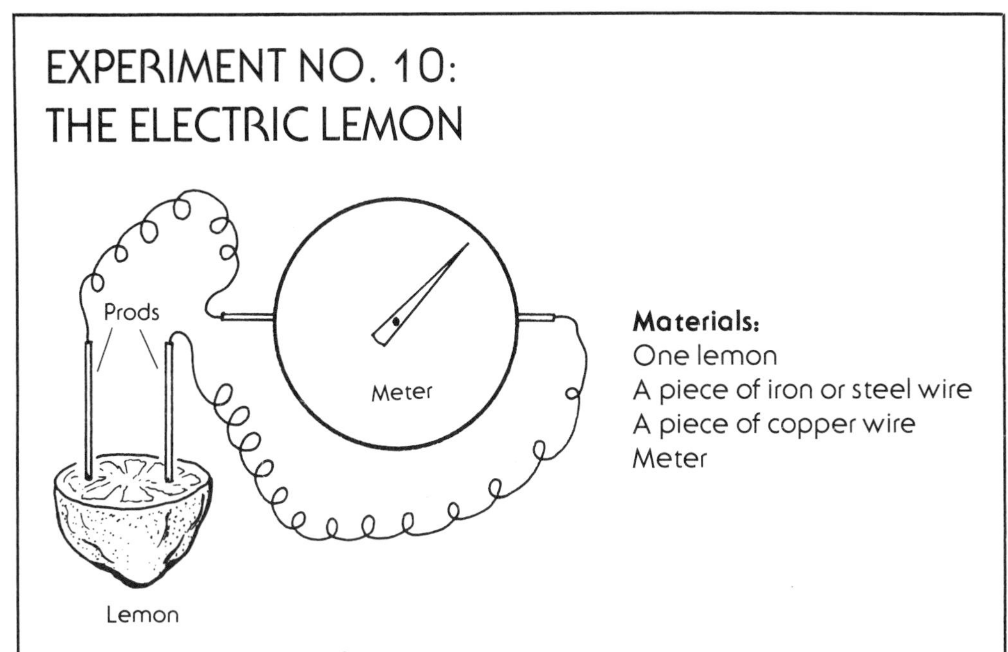

Materials:
One lemon
A piece of iron or steel wire
A piece of copper wire
Meter

Figure 14. The "electric lemon" experiment.

First, strip about an inch of insulation from each of the wires. Roll the lemon on a hard surface to make it as juicy as possible. Cut off a slice from the top. Insert the two pieces of wire into the lemon. Attach the iron or steel wire to the positive (+) side of the meter, and the copper to the negative (−) side.

Does the needle move? Reverse the wires so the copper is on the positive side and the iron is on the negative side of the meter. Do you get movement from the needle?

It's obvious that one arrangement gives you electricity and the reversed position does not. The reason is that the electrochemical action forces the electrons to flow from one metal to the other. It is up to you to determine which metal should be connected to the positive side of the meter to obtain electricity.

Other combinations of various metals are possible besides the one your tried. Try as many different combinations as you can. Don't forget to reverse each combination.

You can list your results by making a table like the one on page 26. This is the way an experimenter works. You avoid doing the same thing twice. Above all, you learn from the tabulated results.

Metal Combinations	Current Flow		
	None	Little	Much

7

The Marriage of Electricity and Magnetism

Coulomb's Measurements

Charles de Coulomb (1736–1807) was the first to measure the amount of electricity and magnetism generated in a circuit. Until then, the flow of electricity could be detected, but not the amount.

We have all sorts of meters nowadays, some that measure as little current as 2 microamperes (a microampere is one millionth of an ampere) or voltages down to 25 millivolts (twenty-five thousandths of a volt). And these are not unusual. But in those days, even the most serious experimenter had no such meters.

There were some crude attempts to measure electricity. Gilbert's versorium could indicate the strength of the magnetic force of an object, but it was not very precise. Coulomb invented the torsion balance. A complicated piece of hardware, it measured very accurately the magnetic force in a circuit. From its use he evolved two important laws: (1) The force between two magnetic poles is in direct ratio to the product of the strength of the two poles. More simply, it means that you multiply the strength of one pole by the strength of the other to find the magnetic force. (2) As the distance between the two poles is increased, the reduction in force between them is equal to the ratio of the square of the shorter distance to the square of the longer distance. For example, two poles are three inches apart and you move them so they are four inches apart. Square each number. 3 × 3 = 9. 4 × 4 = 16. 9 is just a little more than half of 16. So the strength is reduced by about half.

Coulomb also invented an instrument that measured the strength of the magnetic field about the earth. It was known that the earth was a huge magnet with a North Pole and a South Pole. Coulomb's *magnetometer* was able to indicate the exact force of that magnetic field.

The unit of electrical charge is called a *coulomb* in his honor. An electrical charge is the quantity of electrical energy stored in a battery, capacitor, or in any insulated object that is able to hold that energy for a time.

Electricity = Magnetism?

Volta's discovery spurred many inventors. They built larger and more efficient batteries. In some, sulfuric acid was the liquid. Copper and zinc were the favorite metals, but gold and other rare metals were also employed.

The newer batteries found other uses besides the production of electricity. A pupil of Volta discovered electroplating. This system uses electricity to take a metal out of a solution and deposits it on a metal object in the solution. Silver and gold plating are done this way today.

Electricity from batteries was used to isolate

EXPERIMENT NO. 11: THE BATTERY-POWERED COMPASS

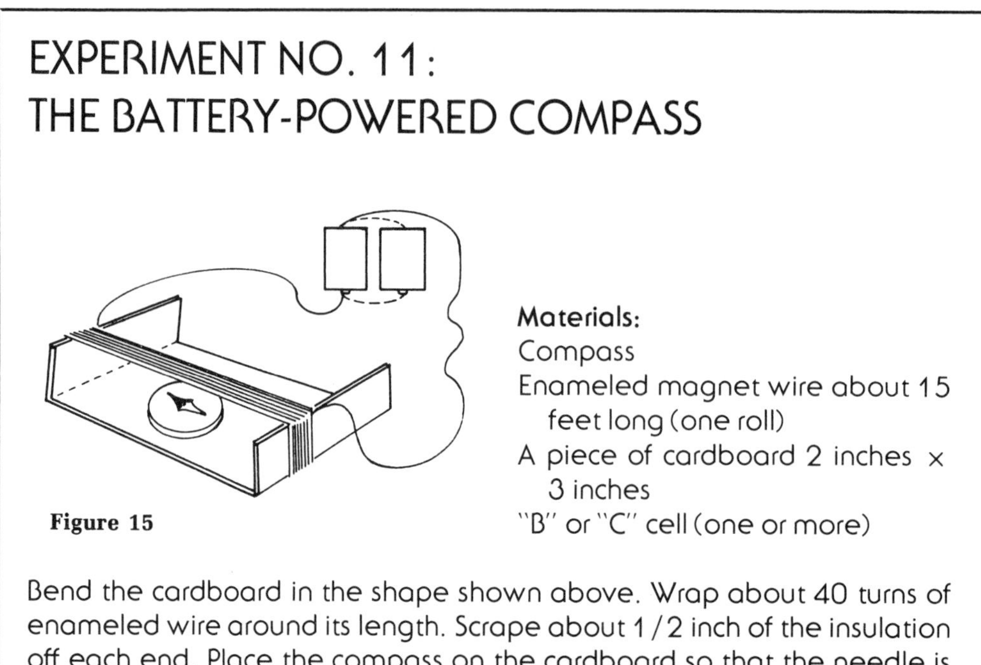

Figure 15

Materials:
Compass
Enameled magnet wire about 15 feet long (one roll)
A piece of cardboard 2 inches × 3 inches
"B" or "C" cell (one or more)

Bend the cardboard in the shape shown above. Wrap about 40 turns of enameled wire around its length. Scrape about 1/2 inch of the insulation off each end. Place the compass on the cardboard so that the needle is at right angles to the wire coil. Connect the wire ends to the cell. The needle will swing back and forth as it is affected by the electric current. Now add one more cell. Note how much more the needle swings. It proves that the amount of needle movement depends on the amount of current passing through the wire. From this fact a current meter—an ammeter—can be built that will indicate how much current is flowing through a wire. The more current, the greater the movement of the meter's needle.

Important note: Since you want more current, the cells must be connected in **parallel.** That is, the negative ends and the positive ends of both cells are connected together. If you were to connect positive to negative, you would get **twice as much voltage,** but the **current would not increase.** Try both arrangements and see the effect on the compass needle.

Now turn the compass around so the needle lies parallel to the winding. Does the current affect the needle?

Make certain when you are trying these experiments with a compass needle that you do not bring the compass close to a piece of iron or steel. The needle would then swing in the direction of the metallic object and throw off your experiment.

elements such as sodium, barium, boron, and potassium, as well as iodine, chlorine, and fluorine.

Magnetism and electricity were still regarded as two separate wonders. Steel had been magnetized by a bolt of lightning and by the use of electricity. Other experiments showed that a piece of steel, such as a needle, could also be magnetized with a friction machine.

Even 20 years after the invention of the battery, a physics professor at the University of Copenhagen, Hans Christian Oersted (1777–1851), treated electricity as a separate entity.

In 1820 he was doing an experiment in front

EXPERIMENT NO. 12: THROUGH THE EYE OF A HELIX

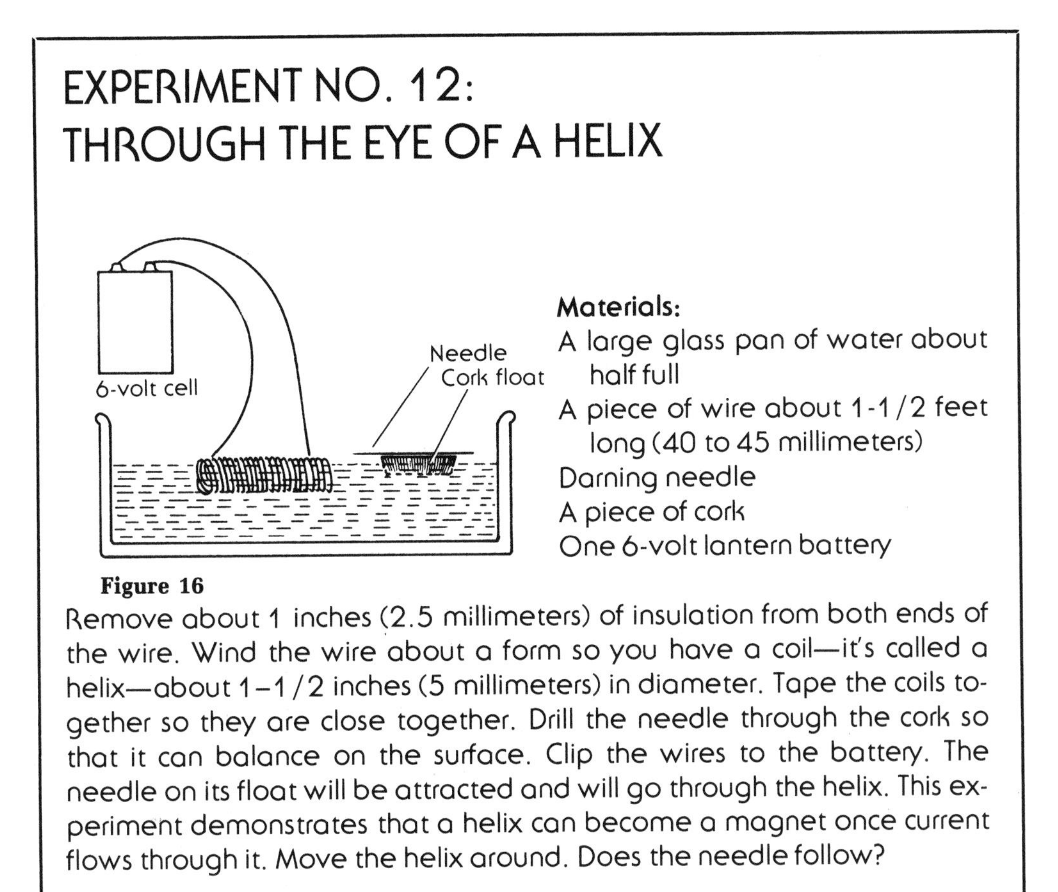

Materials:
A large glass pan of water about half full
A piece of wire about 1-1/2 feet long (40 to 45 millimeters)
Darning needle
A piece of cork
One 6-volt lantern battery

Figure 16

Remove about 1 inches (2.5 millimeters) of insulation from both ends of the wire. Wind the wire about a form so you have a coil—it's called a helix—about 1-1/2 inches (5 millimeters) in diameter. Tape the coils together so they are close together. Drill the needle through the cork so that it can balance on the surface. Clip the wires to the battery. The needle on its float will be attracted and will go through the helix. This experiment demonstrates that a helix can become a magnet once current flows through it. Move the helix around. Does the needle follow?

of his class showing that electricity could be converted into heat. (We use that effect today in our toasters and electric heaters.) Oersted placed a compass close to the electric circuit lying on the desk used for the demonstration. He happened to place the compass *parallel* to the wire lying there. As he switched the circuit on, he saw something strange. To his surprise, the compass needle swung violently back and forth as he opened and closed the circuit.

The oscillation (rapid movement) of the needle puzzled him. Fascinated with the needle's movement, he experimented with stronger batteries that provided him with *more current*. As he increased the current, the needle's oscillations increased in proportion. More experiments showed that a wire with electricity running through it deflected a compass.

From these experiments, Oersted showed for the first time that electricity and magnetism were similar. Scientists everywhere tried his experiments with variations to check his results. There was no doubt about the importance of what he discovered: Electricity flowing through a wire could deflect a compass just as a magnet could. There were far-reaching results from Oersted's discovery. It is possible to say that his experiment showing the effect of electricity on a magnet was the first step toward the production of electricity on a large scale—enough electricity to light the world.

Experiment No. 11 (Fig. 15) gives you the chance to learn what Oersted did. Experiment No. 12 (Fig. 16) is a fun variation on Experiment No. 11.

Electromagnetism

André Ampère was a physicist and mathematician. He was born in Lyons, France, in 1775 and died in 1836. He contributed to the invention of the galvanometer as a means of accurately measuring current. He also had the idea that electromagnetism could be used for te-

EXPERIMENT NO. 13: CHECKING UP ON ARAGO

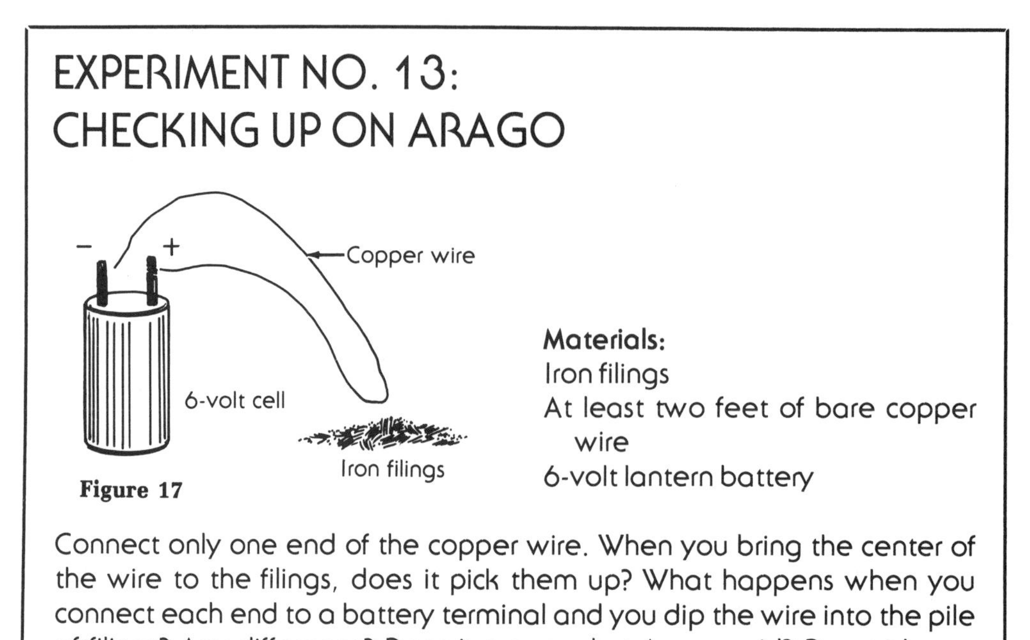

Figure 17

Materials:
Iron filings
At least two feet of bare copper wire
6-volt lantern battery

Connect only one end of the copper wire. When you bring the center of the wire to the filings, does it pick them up? What happens when you connect each end to a battery terminal and you dip the wire into the pile of filings? Any difference? Does it prove what Arago said? Do not leave the wire connected to both terminals for long—you will short the battery out. A few seconds should prove whether Arago was right or not.

legraphy, but it was not until 1835 that Samuel Morse put that idea to work.

Through Ampère's efforts, electromagnetism became more than a scientific curiosity. He found that two wires close to each other reacted in a strange fashion if one wire had current flowing through it.

During 1820 he described his experiments to the French Academy, the foremost scientific society in France at that time. He showed that a force existed that diverted a compass needle when current passed through a wire close to the compass. (You have already seen this in Experiment No. 11.) But he also proved that the needle would indicate which way the direct current was flowing.

François Arago was another French physicist (1786–1853). He was also an astronomer, and he had many other interests, including calculating the speed of sound. In those days, it was quite common for a scientist to be involved with many sciences. Little was known about many of them, and so it was relatively easy to become an expert. Arago is of interest to us because of his experiments in the field of electromagnetism.

You already know that a magnet attracts iron filings. In 1820 Arago proved that a wire carrying current will do the same—*but only as long as the current flows.* Arago's experiments proved that electricity and magnetism exhibited some characteristics in common.

The work of Arago and Ampère led to the discovery that a soft iron bar could be magnetized just like the wire. A loop of wire passing around the bar will magnetize it while the current is flowing through the loop.

Experiment No. 13 lets you test Arago's conclusions (Fig. 17), and Experiment No. 14 shows you how to build an electromagnet (Fig. 18).

Electromagnets, Motors, Relays, and Solenoids

The electromagnet is used industrially where heavy masses of steel or iron must be lifted and carried. At the end of a cable, a huge plate made of iron is suspended from a crane. When the current is turned on, the plate—which is a powerful electromagnet—is magnetized and

EXPERIMENT NO. 14: MAKING AN ELECTROMAGNET

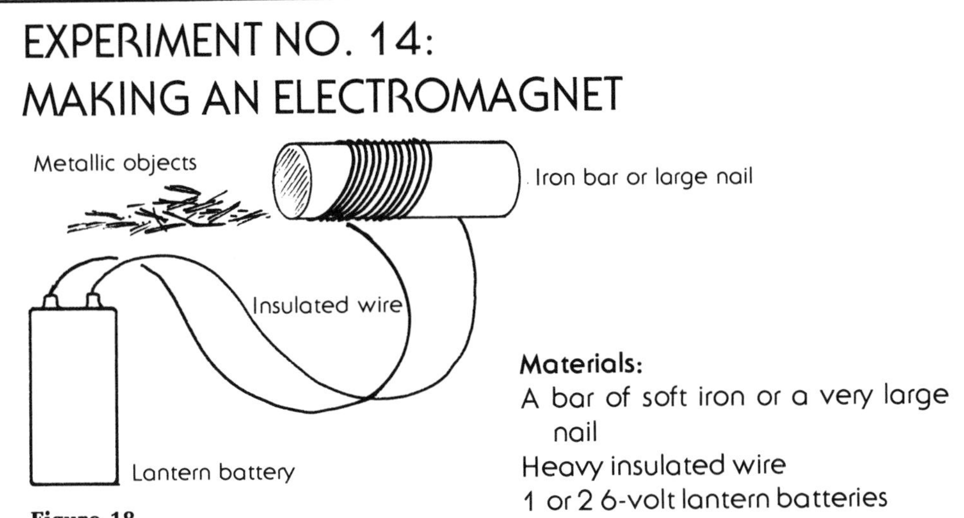

Figure 18

Materials:
A bar of soft iron or a very large nail
Heavy insulated wire
1 or 2 6-volt lantern batteries

Wind about 20 turns of insulated wire around the iron bar. The diameter of the coil of wire should be less than half of its length. Leave long leads. Remove about an inch of insulation from each end. Get a bunch of iron or steel nuts and bolts. Any metallic objects that will be attracted to a magnet will do. Connect the wires to the one battery. Does the iron bar act as a magnet?

Lay the bar down on your work area and see how far a bolt can be and still be drawn to the electromagnet. Measure that distance. Disconnect the battery.

Instead of 20 turns, wrap the wire 40 times around the iron bar. Reconnect the battery. Is there more attraction than with only 20 turns? Measure the distance across which the bolt will be drawn to the magnet.

Now add another battery in series to the first. Again measure the farthest distance between the bolt and the magnet. How much difference than when you had only one battery?

Now repeat with only 20 turns of wire but with both batteries. How does this compare with the 40 turns of wire? Write down the results so you have a comparison, and you can then judge the effect of more windings and/or more current.

When you disconnect the batteries from the electromagnet, does the bolt fall off at once? Usually it will hang on for a few seconds. This depends on the type of iron that makes up the bar. The remainder of electromagnetism is called **residual** magnetism.

becomes capable of lifting many times its own weight. Tons of material are transported from one place to another without needing to tie cables around each load of steel. When the current is turned off, the steel drops off.

William Sturgeon (1783–1850) is credited with having built the first electromagnet in 1821. He is also supposed to be the first to use an iron core for the magnet. Iron, particularly soft iron, will magnetize easier and will lift a greater weight for its size than will steel (see Fig. 19).

Sturgeon was a bootmaker. He enjoyed building electrical apparatuses, and he gave many lectures with the equipment he built. Not trained as a scientist, he had few theoretical

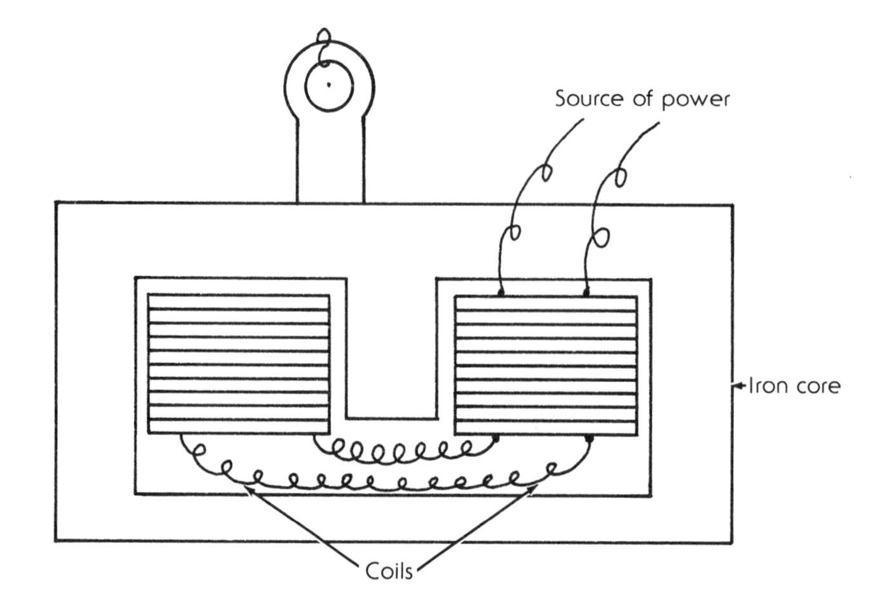

Figure 19. An electromagnet.

ideas. A practical man, he probably did not know or care why something worked. His question was, "Will it do what it's supposed to?"

Sturgeon is also credited with having invented the first working electric motor. (Unfortunately, few records of his work survive.) Another invention of his was the commutator—the part of an electrical generator that collects current by means of brushes from the revolving rotor.

A *relay* is also the result of electricity and magnetism operating in one unit. A coil of wire is wound around a soft iron bar (see Fig. 20). When the current flows through the circuit, the bar is magnetized and pulls in a strip of metal that closes a circuit. In this way, a tiny current can control another circuit—close it or open it—that contains many amperes of current. The relay can also control a circuit remotely. Experiment No. 15 shows how to build a simple relay.

The *solenoid* (Fig. 21) is another example of

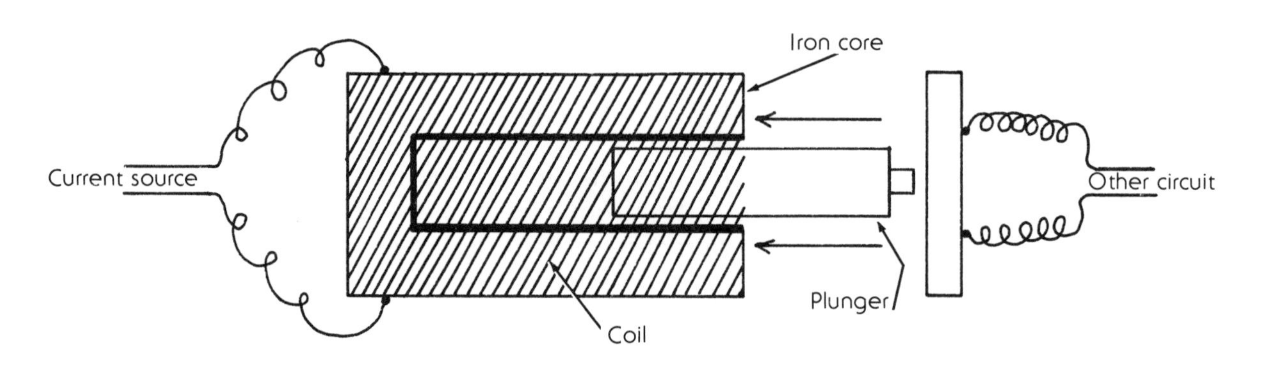

Figure 21. A solenoid.

EXPERIMENT NO. 15: A SIMPLE RELAY

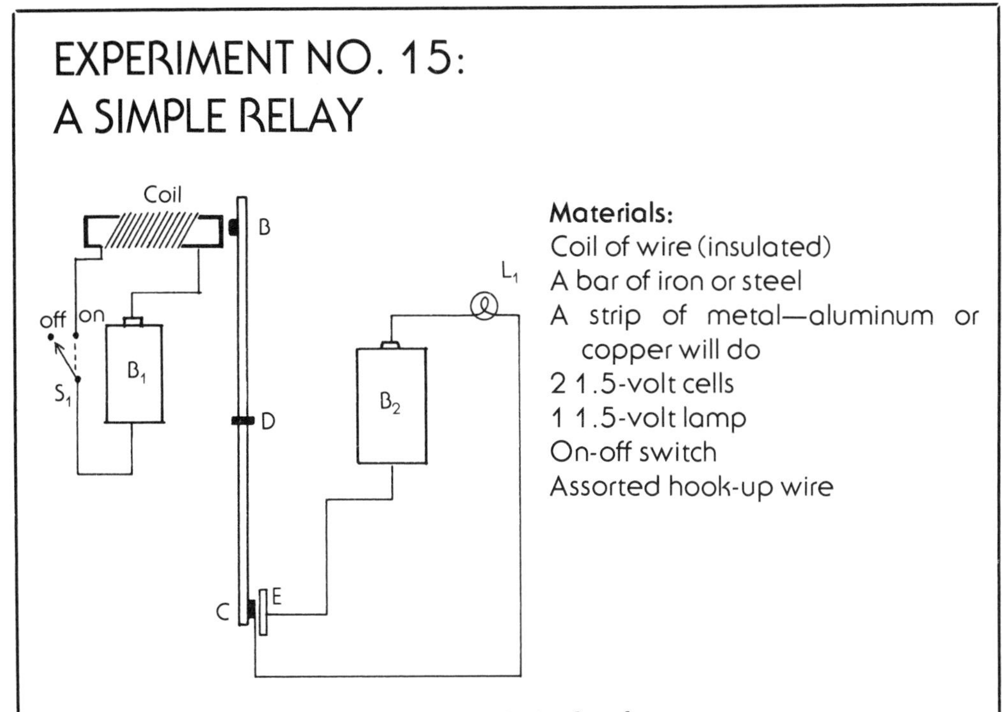

Materials:
Coil of wire (insulated)
A bar of iron or steel
A strip of metal—aluminum or copper will do
2 1.5-volt cells
1 1.5-volt lamp
On-off switch
Assorted hook-up wire

Figure 20. A simple relay.

A coil of insulated wire is wound around the metal bar. When the switch S_1 is closed, current from cell B_1 flows through the coil and turns the bar into a magnet. This pulls in metal strip **B–C** so that Point **B** makes contact with the magnetized bar. The metal strip pivots at Point **D** so that **C** touches **E. B** and **C** must be insulated from each other. The second circuit is now closed. In the second circuit in this case is cell B_2 that lights the lamp L_1 when the circuit is closed.

It requires only a few volts and a few milliamps to cause the coil to pull in **B.** When the switch S_1 is opened, the coil loses its power to hold the metal strip against the iron bar, and the lamp will go out.

By these means, a tiny voltage can control a much larger voltage in the second circuit. High voltages are often controlled by relays that eliminate high voltages running through long wires.

A relay can also be controlled automatically by a clock or by a thermostat or even by water rising above a certain level. This would immediately start a pump in the second circuit.

magnetism and electricity working together. When the current flows through the coil, the plunger is pulled in. The iron core moves into the coil where the magnetic force is the greatest. Solenoids are used to break a circuit—just like a switch—or to complete one, depending on how the solenoid is designed.

8

The Heat Is On

It was mentioned previously that electricity can be converted to light and heat. A glowing light bulb is one example of the conversion of electricity to light. A toaster changes electricity to heat.

What about the reverse? Can you convert heat into electricity? These were the questions that Thomas Seebeck (1770–1831), a German physicist, asked himself. To find the answer, he twisted two wires made of different metals and heated the junction where the two wires met. He produced a small current. This process is called *thermoelectricity (thermo* is a form of the Greek word meaning "heat." So thermoelectricity means electricity generated by means of heat.) Later, we'll learn about other conversions—electricity converted into heat and light converted into electricity.

The junction of two dissimilar wires is called a *thermocouple.* When heat is applied to a thermocouple, a voltage difference is created between one metal and the other. This difference shows up at the free ends of the wires. The more heat you apply, the more voltage you get. (Until, of course, you apply too much heat and the wires melt.) Experiment No. 16 shows you how the thermocouple works.

Could you heat water by using thermo junction heated by the sun? Unfortunately, the wires won't develop enough current to raise the temperature of the water so it is noticeable.

Some industrial processes use this priniciple to measure temperature. Two prods made of different metals are dipped into the solution to be measured. As the solution heats up, the voltage increases. A meter is connected to the ends of the wires just as in the experiment above. This meter has one difference. It is calibrated in degrees—usually Celsius. (On the Celsius scale, water freezes at zero degrees and boils at 100 degrees.)

The person in charge looks at the meter to see when the right temperature is reached. He or she can stop the heat manually, but most of the time it is done automatically. As the needle on the meter reaches a preset point, the heating circuit is turned off.

As you realize, all of this is based on a very simple idea. But even the simplest idea had to be thought of by someone.

Seebeck showed that a heated thermocouple produces electricity. What about reversing the process? What would happen? A French watchmaker, Jean Charles Peltier (1788–1842), experimented further with Seebeck's discovery. He passed a weak electric current through the junction of two different metals. He found that a cooling effect took place if the current went in one direction. If the current was reversed, heat was generated. This is now known as the *Peltier effect.*

A point to keep in mind is that an idea can often be extended into a new discovery. An inventor is often someone who takes someone else's concept and tries different things with it. They don't all work, and all inventions are not necessarily great. But to experiment is to try. And you can learn from your failures.

EXPERIMENT NO. 16: ELECTRICITY BY THERMOCOUPLES

Materials:

About 3 feet of heavy copper wire
Same length of heavy gauge iron wire
Candle
Current meter

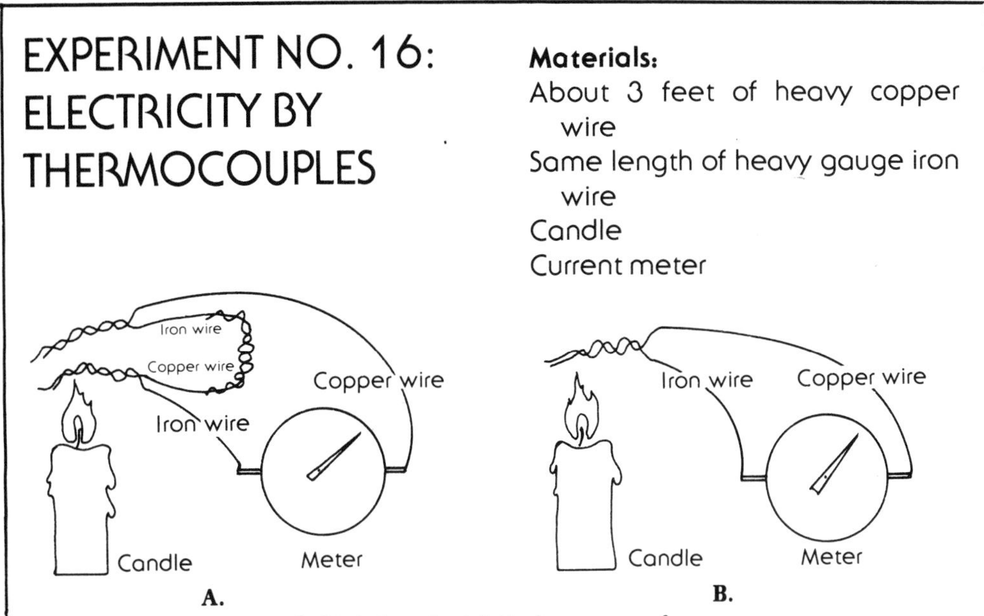

Figure 22. Two ways of obtaining electricity by means of thermocouples.

Here's a simple illustration of thermoelectricity: Cut one length of copper wire and one of iron, each 12 inches long. Twist the wires very tightly together for an inch or so. Only the ends need to have their insulation removed. The other ends are connected to a store-bought meter or the current indicator you built in Experiment No. 11 (see Fig. 22A).

Make sure to hold the wires over the heat with pliers in order to avoid burning yourself. As the wires become hot, you will see a tiny increase in current on your meter or indicator. You can try other metals joined together and see which combination produces the most current.

The amount of electricity generated will be small. If you want to get a bigger reading—more current—attach two sets of wires as shown in Figure 22B. The two thermal junctions will be in parallel. This will generate more current.

A way of obtaining free electricity is to use the sun to heat the wires. To make the experiment work, you need a bright sunny day with no clouds. To get more current from your thermo junction, use a magnifying glass so the sun is focused on the twisted wires. Put some aluminum foil or a mirror under the wire to reflect the heat. Do not let the foil touch wires—the electricity would be conducted away, and you would have no method of judging the results. Your reading on the ammeter should be greater than the reading you obtained with the candle.

9

Laying the Groundwork for City Lights

Faraday: Electrical Pioneer

The English physicist Michael Faraday (1791–1867) combined what had been known about electricity and magnetism with a few ideas of his own. It is that combination that forms the foundation of our present day technology of generating electricity on a large-scale basis.

Faraday was originally an instrument maker. He was assistant to Humphry Davy, who was superintendent of the Royal Institution of London. When Faraday was only 29, he took over for Davy when the latter retired.

He began to do a series of experiments having to do with the relationship between electricity and magnetism. His pioneering work dealt with understanding how electric currents work. From his many experiments, many practical inventions would come, such as the motor, generator, transformer, telegraph, and telephone. But they would come fifty to a hundred years later. He created words such as *electrode, anode, cathode,* and *ion* to describe his work. We still use these terms today in electricity and in electronics.

Let's take a look at a few of his experiments. You can perform Experiment No. 17 (see Fig. 23) and see what he found out when he showed it to the scientists of his time.

What do you learn from Experiment No. 17? Let's analyze what went on. As you know, there is no connection between coil *A* and coil *B.* When you close the switch, current flows from the battery connected to coil *B.* This current flow builds up a magnetic field cutting across coil *A.* The needle of the meter moves, indicating that an electromotive force *(emf)* is *induced* in coil *A.* Once the magnetic field is built up, the needle goes back to zero.

Open the switch. What happens? The needle jumps up for a moment and goes back to rest. The reason is that there is no current through coil *B.* So the magnetic field collapses. As it collapses, it cuts across coil *A.* Once again an emf is generated. This time the lines of force cut across the coil, but in the reverse direction. The emf flows in the opposite direction in coil *A.* The needle is deflected in the direction that is the reverse from the one it showed when you first closed the switch.

Faraday's experiment on inductive current is a very important one, as you will see as we go along.

What's the difference between *conduction* and *induction?* By now you should know the answer. Current is *conducted* from one part of a circuit to another part *only when there is a physical connection between the parts.* The connection can be a wire, a switch, or any component that makes the connection.

Induction is when a voltage or current is produced by lines of force. And there is *no* physical connection.

One further fact: The current in the induced circuit (coil *A*) always flows in the direction which is the reverse of the current flow in the electrified circuit (coil *B*).

EXPERIMENT NO. 17: SHOWING INDUCTIVE CURRENT

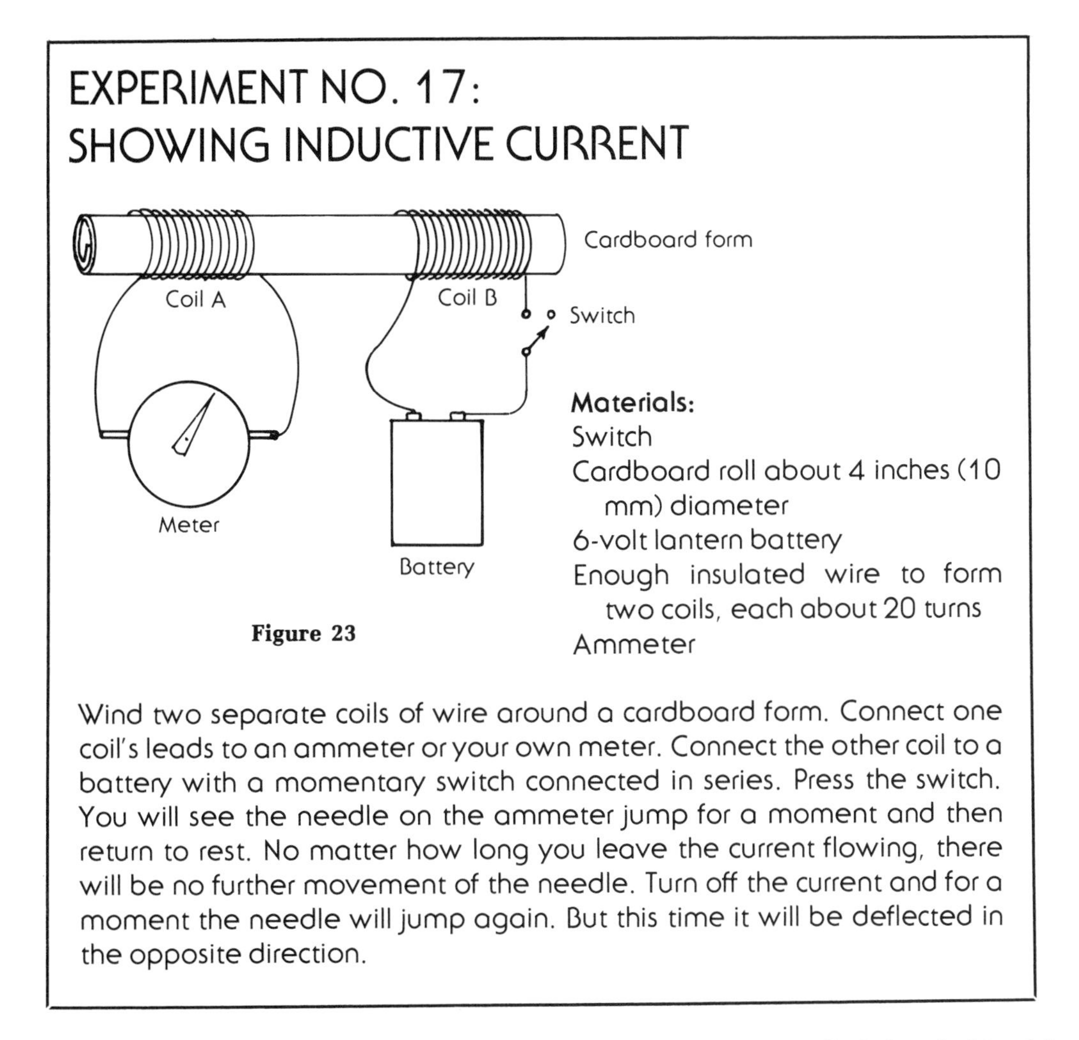

Figure 23

Materials:
Switch
Cardboard roll about 4 inches (10 mm) diameter
6-volt lantern battery
Enough insulated wire to form two coils, each about 20 turns
Ammeter

Wind two separate coils of wire around a cardboard form. Connect one coil's leads to an ammeter or your own meter. Connect the other coil to a battery with a momentary switch connected in series. Press the switch. You will see the needle on the ammeter jump for a moment and then return to rest. No matter how long you leave the current flowing, there will be no further movement of the needle. Turn off the current and for a moment the needle will jump again. But this time it will be deflected in the opposite direction.

Michael Faraday was a different experimenter from many others you've been studying. Not only did he prove his theories right or wrong, but he would examine the side issues yielded by his experiment. Often he came up with a new idea based on an observation that could be far removed from what he had been looking for. Sometimes such an experiment showed him his original theory was wrong.

This never discouraged him. He would say, "The failures are just as important as the successes." He felt—and he was right—that failures teach just as do the successes.

Experiment No. 18 enables you to explore another Faraday device, the electric generator (Fig. 24).

Remember Experiment No. 4, in which you dropped iron filings on a piece of paper laid over a magnet? The pattern in which the filings fell into place was called the "fields of force." Faraday invented that term in 1831 to explain the action of the whirling disk. What the disk was doing was cutting across the magnet's fields of force. This action produced a current that lasted as long as the crank was turned.

You could produce a steady current if you had a water wheel in a fast moving stream. The shaft of the wheel would also be the shaft of the disk. As the wheel spun under the force of the water, the disk would turn and presto!—electricity. But Faraday was not interested in the mechanics of electricity, only the principles. What really interested him was the interaction of magnetism and electricity as represented by coils of wires or whirling disks. But he had more tricks up his sleeve.

In 1832, one year after showing the experiment on inductive current, Faraday read a paper to the Royal Society in which he showed

EXPERIMENT NO. 18: FARADAY'S ELECTRIC GENERATOR

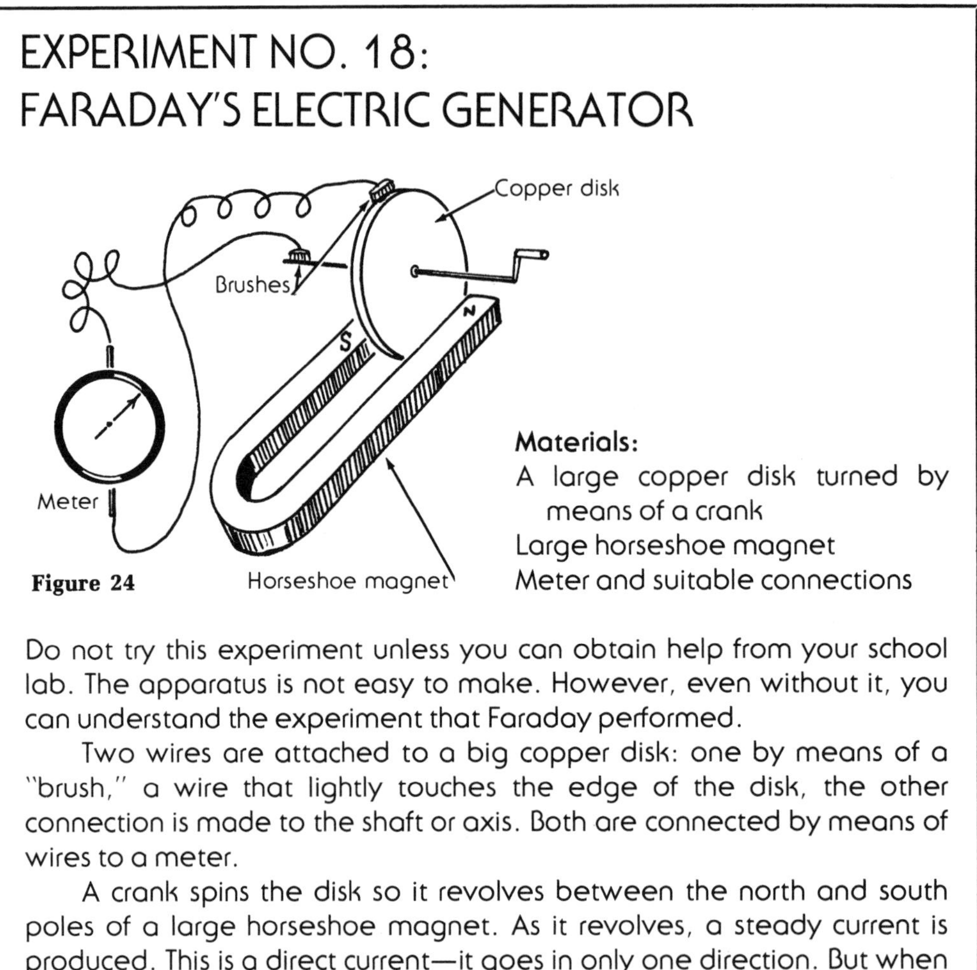

Figure 24

Materials:
A large copper disk turned by means of a crank
Large horseshoe magnet
Meter and suitable connections

Do not try this experiment unless you can obtain help from your school lab. The apparatus is not easy to make. However, even without it, you can understand the experiment that Faraday performed.

Two wires are attached to a big copper disk: one by means of a "brush," a wire that lightly touches the edge of the disk, the other connection is made to the shaft or axis. Both are connected by means of wires to a meter.

A crank spins the disk so it revolves between the north and south poles of a large horseshoe magnet. As it revolves, a steady current is produced. This is a direct current—it goes in only one direction. But when the crank is turned in the opposite direction, the current also reverses direction, as the meter indicates.

that the earth itself was a huge magnet—which, like a magnet, produced lines of force. He proved his theory simply by using the familiar helix (coil of wire) and a magnet. He had none of the astounding tools that men of science work with today. His most important tool was his brain.

But let's look at another Faraday experiment. He showed that an *extra current is induced in the primary coil* at the moment the circuit is opened and closed. This is known as *self-induction* (see Fig. 25).

When the switch is opened, current flow is stopped in the circuit. The *self-induced voltage* opposes this drop. The voltage tries to build up a current in the opposite direction. To do that, it must leap across the air gap in the open switch. A spark jumps across the gap. The

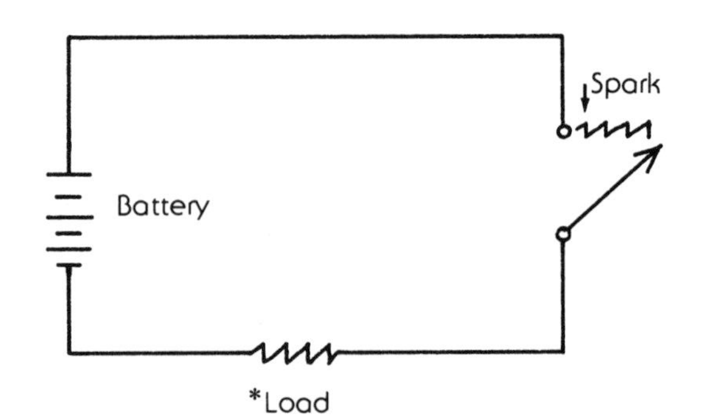

Figure 25. Self-induction. *Load means anything in the circuit that will be consuming part of the voltage. This could be a very thick wire or an electric light bulb. In some instances, when the spark has enough power, it will end up by burning a hole in the switch just at the point where the spark leaps the gap.

length of the spark depends on the amount of voltage produced by the battery.

The spark occurs only at the instant that the current changes to the other current. If you leave the switch permanently open, no spark occurs. This means that when there is a steady current flowing, there can be *no* self-induction.

AC-DC

So far you have been dealing with direct current and voltage as supplied by a battery. Direct current (DC) flows only in one direction. It leaves the negative battery terminal, goes through the circuit, and returns to the positive terminal.

A coil inducing current in another without any connection is the beginning of a transformer. The basic difference is that the transformer operates with alternating current (AC).

AC changes direction continually. As you remember from your last experiment, current was induced only when the flow in the electrified circuit was opened and closed. Since the direction of the current (AC) changes, it induces a current that also changes but is constant. In the North American continent, the voltage supplied in most homes changes direction 120 times a second. To complete one cycle takes two changes, and so it is called 60-hertz AC. One hertz equals one cycle per second.

We'll return to alternating current later. For the time being, we have to work with the only voltage available to you—direct current from a battery.

Laying the Groundwork for City Lights

Let's return to Michael Faraday. He had made yet another major contribution to the science of electricity—the knowledge necessary to generate electricity on a large scale. (If he found a way to do that, why didn't he build a generating plant and electrify England back then? Faraday was simply interested in knowing the "why" of some electrical and magnetic behavior. He left the actual use of his theories to those who came later.)

Faraday's principle for generating electricity was the same principle we use today in an electric utility plant. To see what Faraday did, take a look at Experiment No. 19 (see Fig. 26).

Now that you've either read or performed Experiment No. 19, you're probably saying, "That's all right as far as it goes. But that isn't enough electricity to light up a flashlight bulb. How do you go about generating enough electricity for an entire city?"

You have the clue—more wire or a larger magnet. The principle laid down by Faraday is still right.

Electricity is generated in a large utility plant by means of a three-step process. Water is heated until it turns to steam. The steam expands and rushes out of the boiler through a pipe that acts like a jet. The force of the stream of steam hits the vanes of a *rotor*. The vanes are winglike projections on a huge coil of wire. The coil of wire spins, due to the force of the steam.

The moving coil is called a rotor because it rotates inside a huge magnet. The magnet, being stationary, is called a *stator*. This combination of rotor-stator is called a generator. The faster it spins and the more turns of wire are in the rotor, the more voltage will be generated.

The rotor inside the stator in a utility plant is several stories high. Regardless of its size, it illustrates Faraday's principle—interrupting the field of force around a magnet will produce electricity.

You will notice that I never used the word *create*. It is because a magnet has a field of force around it that we can *generate* elec-

EXPERIMENT NO. 19: PROVING FARADAY'S THEORY

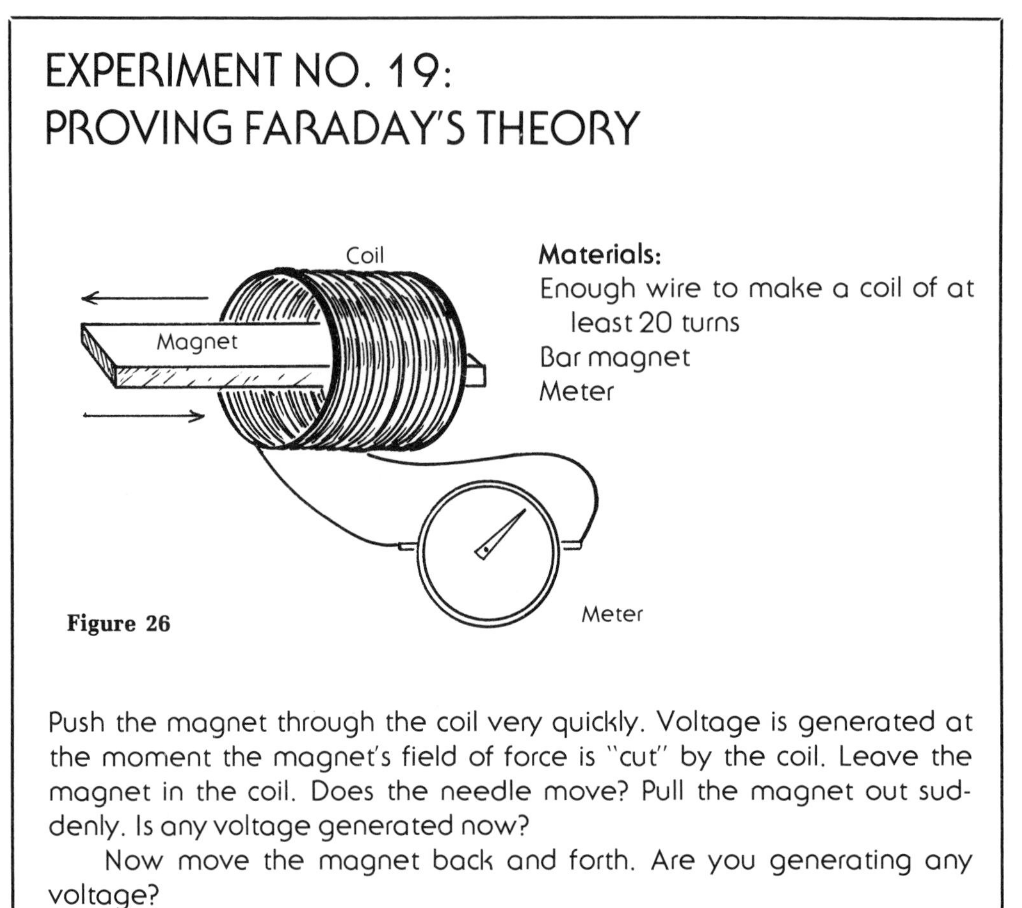

Figure 26

Materials:
Enough wire to make a coil of at least 20 turns
Bar magnet
Meter

Push the magnet through the coil very quickly. Voltage is generated at the moment the magnet's field of force is "cut" by the coil. Leave the magnet in the coil. Does the needle move? Pull the magnet out suddenly. Is any voltage generated now?

Now move the magnet back and forth. Are you generating any voltage?

From this experiment you can see that voltage is generated only when the field of force is interrupted by the coil. This is true whether the coil moves or the magnet moves. When both are at rest, there is no voltage.

With some 20 turns of wire, you will notice that the needle on the meter moves only slightly. If you were to coil a whole spool of wire and push the magnet in and out, the needle would really jump. This shows that the bigger the coil, the more voltage will be generated. A larger magnet will also produce more voltage than a smaller one.

Turn the magnet around so the poles are reversed. Any difference in the way the needle moves? Why?

tricity. Without that field of force, we would not have any voltage. But we are not *creating* electricity. This is an important point. It is so important that it has been incorporated into the all-encompassing Law of Thermodynamics, which states that matter can neither be created nor destroyed. For example, you may ask, "When a house burns to the ground, isn't it destroyed?" The house as a dwelling place is destroyed. That's true. But the basic matter of which the house is composed is not destroyed. Part of the house has been converted into ashes, and part into heat that has escaped into the air. But the matter is still there somewhere. Even if we can't see it, touch it, or smell it.

Let's say you have a 6-volt battery in a circuit that contains a tiny light and a small motor. If three volts are used up in this circuit, are they destroyed? No. Part of the voltage was changed into heat, light, and movement, but it is still present. Matter can change form, but it is never destroyed. Think about other examples. You will always find the same answer.

In honor of his great contributions to the science of electricity, Michael Faraday's name is linked to the unit of capacitance. It is called the *farad*. Capacitance measures the electrical capacity of a capacitor.

From Induction to Magnetodynamics with Joseph Henry

Honored all over the world for his contributions to the advancement of electricity, Joseph Henry (1799–1878) is almost unknown in his own country, America. He was the first director of the Smithsonian Institution in Washington, D.C., and was a teacher of physics. He discovered (or rather he rediscovered) the principle of electromagnetic induction. He did not know that Faraday had already come to the same conclusions. When he found out, he was quick to give the Englishman full credit for the discovery.

On the other hand, Joseph Henry discovered self-induction ahead of Faraday. In his honor, the unit of inductance is called the *henry*. The inductance in a circuit equals one henry under a current charge of one ampere per second, inducing one volt.

Henry's work in electromagnetism is important because he showed the possibility of an electromagnetic telegraph. Later inventors, such as Morse and Edison, were to add the refinements that make the telegraph the useful instrument we know today.

Henry constructed his electromagnet in a slightly different way. First he wound a coil of wire around an iron core. Then he wrapped another coil over the first. Each coil added to the lifting power of the electromagnet. (Wire in those days was not normally insulated. Obviously they couldn't wrap a coil over another without insulation, so they would wrap silk or cotton around each wire.)

When Faraday pushed a magnet back and forth through a coil, he had no idea that much of the world would be using his discovery to generate electricity on a large scale.

His discovery led to yet another way of producing electricity directly, called *magnetohydrodynamics*. The word *magnet* in the name may give you some idea of what to expect.

Here is how the process works: A hot gas is formed by burning coal and passing water over it to produce steam. This won't conduct electricity, so the steam is made to come in contact with a potassium salt. The resulting conductive gas is forced to pass across a magnetic field, and reacts the same way as in the experiment you just did. Electricity is generated.

The process is "clean" in that there are few wastes to pollute the air. It is still in the experimental stage, but great things are expected from it.

This is a further example of how a simple experiment can lead to spectacular ends.

Gauss, Man of Units

Three scientists of this period who made significant contributions are Karl Gauss, James Clerk Maxwell, and H. E. Lenz.

Karl Gauss (1777-1855) did not discover any electric or electromagnetic principles. What Gauss did was to create a set of units to measure the amount of magnetic induction. A unit of magnetic induction is called a *gauss*.

Until 1832 there was no way to measure magnetism. Think what would happen if we had no system for measuring electricity! If you plugged in a radio to a battery, you could blow it up because you would not know how much voltage the battery could provide and how much was needed to run the radio.

When a reel of magnetic tape—audio or video—is demagnetized, we use a *degausser* to do the job. A TV set has a coil across the picture tube that automatically degausses it so no residual magnetism remains to spoil the picture.

Maxwell's Mathematics

Back we go across the English Channel to visit James Clerk Maxwell (1831–1879). From early youth, he was interested in scientific ideas. He did much work on the theory of gases. At the age of 40, he became a professor of physics at Cambridge.

His contribution to science is that he made many of Faraday's ideas clearer. He put them in mathematical form so they could be better understood. He tried to explain, as did Faraday, the lines of electromagnetic force.

At first it would seem that his work was of little consequence. But it was a step forward—a step that helped scientists who came later. As an example, the Maxwell-Boltzmann law was developed as a result of Maxwell's equations. The law gives the distribution of the speeds among the molecules of a perfect steady-state gas.

A *maxwell* is the electromagnetic unit of magnetic flux. Flux are the electric or magnetic lines of force about a magnet or an electric wire.

Lenz's Law

A Russian scientist named H. E. Lenz (1804–1865) is remembered for finding the law governing the direction of an induced current: The direction of the current induced in a conducting circuit by its motion in a magnetic field is such as to produce an effect opposing the actual motion in the circuit.

Lenz also studied the Peltier effect, mentioned in Chapter 8. Peltier had found that when a weak electric current is passed through the junction of two dissimilar metals, a cooling effect takes place if the current goes in one direction, and that reversing the current results in heat being generated. Lenz took this idea one step further. He showed you could freeze small amounts of water using the Peltier effect.

Ohm's Law

To visit the next scientist, we must jump back to Europe. George Ohm was a German physicist (1787–1854) who discovered the relationship among voltage, current, and resistance. This relationship exists in any circuit employing *direct current*.

So we can understand Ohm's Law, let's define the terms we're going to use. *Voltage* is the force that pushes current through a circuit. Since it is also called electromotive force, its symbol is E. Current, the rate at which electricity moves from one point in a circuit to another, is symbolized by the letter I. And resistance, which is the opposition to the flow of current in a circuit, uses the symbol R.

Ohm not only showed the relationship among the three units, but he also specified them in mathematical terms. *Ohm's Law* states that one volt of electricity is needed to force one ampere of current through one ohm of resistance.

Algebraically, the law reads: $E = I \times R$. You can also turn it around so that it stands like this: $I = E/R$. This means that the voltage

COMMON ELECTRICAL UNITS AND ABBREVIATIONS

Quantity Measured	Unit	Abbreviation
voltage	volt	V or v
current	ampere (amp)	A or a
resistance	ohm	Ω (Greek letter "omega")
capacitance	farad	F or f
inductance	henry	H or h
power	watt	W or w
frequency	hertz	Hz

Prefixes occur often in the terms used for electrical units. Some units with prefixes, like kilowatts and megatons, you may already be familiar with. Abbreviated forms, like kw or pf, are used in technical literature.

Prefix (Abbreviation)	Meaning	Example (Abbreviation)
pico- (p)	one trillionth (0.000000000001 or 1/1,000,000,000,000)	picofarad (pF or pf) = 0.000000000001 farad
nano- (n)	one billionth (0.000000001 or 1/1,000,000,000)	nanofarad (nF or nf) = 0.000000001 farad nanosecond = 0.000000001— second
micro- (μ)	one millionth (0.000001 or 1/1,000,000)	microvolt (μV or μv) = 0.000001 volt
milli- (m)	one thousandth (0.001 or 1/1,000)	milliamp (mA or ma) = 0.001 amp millihenry (mH or mh) = 0.001 henry
kilo- (k or K)	one thousand (1,000)	kilovolt (Kv) = 1,000 volts kilohm (K or k) = 1,000 ohms
mega- (M)	one million (1,000,000)	megawatt (MW) = 1,000,000 watts megohm (M) = 1,000,000 ohms
giga- (G)	one billion (1,000,000,000)	gigahertz (GHz) = 1,000,000,000 hertz

One ohm is a very small quantity, so most resistors—electronic parts that create resistance—are measured in thousands or millions of ohms. Notice that the abbreviated units for large numbers of ohms are **K** or **k** (1,000) and **M** (1,000,000) instead of K Ω and M Ω, as you might expect. On the other hand, one farad is a large amount of capacitance, so large that most capacitors—devices for storing electrical charge—are measured in fractions of farads, often in microfarads and picofarads.

divided by the resistance is equal to the current.

To see if it makes sense, let's compare the flow of electricity to the flow of water through a pipe. Say you have two pipes. One has twice the diameter of the other. Naturally the water will flow at a faster rate through the larger pipe. The reason is that it offers *less resistance* to the flow.

Double the pressure of the water through the smaller pipe and the water will come out at the *same* rate of flow as it did in the larger pipe—remember, the smaller pipe is half the diameter of the larger pipe.

Using Ohm's Law, we also see that if we increase the voltage (the pressure in the circuit), we get a faster flow of amperes. Think about this for a moment. Try it on your kitchen sink. The faucet is turned off. No water flows. That's the same thing as saying there is no voltage. Turn the faucet wide open. A lot of water comes out. Why? You have *reduced the resistance.* Turn the faucet half way off. The resistance is increased so that less water comes out—just as with an electrical circuit.

If you let the same amount of water continue to flow through a very small pipe, less water will come out of the end because you have increased the resistance. It will come out in greater force, but the actual amount of water coming out per minute will be less than when the tiny pipe is *not* in the circuit.

George Ohm stated something more. He said, "The power in an electrical circuit is equal to the current (in amps) multiplied by the voltage."

Algebraically, he wrote it this way: $P = I \times E$. (P is the symbol for power. The unit for power is a watt.) What is Ohm saying? If you increase the voltage or the current or both, you increase the power. Furthermore, if you increase the pressure—of water or electricity—you will get more power in your circuit. Look at Figure 27A. What it represents is that if you increase the force by two, it is the same as decreasing the resistance by one half.

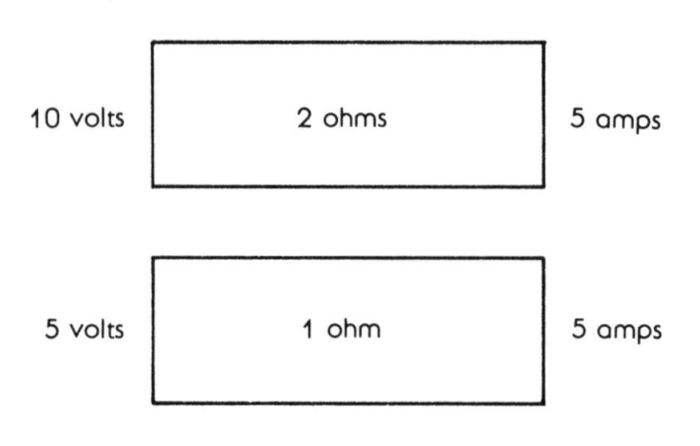

Figure 27A

To help you memorize Ohm's Law, the following diagram (Fig. 27B) will be of help. It states the law in three different ways: $E = I \times R$; $I = E/R$; and $R = E/I$. Simple as these equations appear, they are of utmost value when you are dealing with a circuit in which *direct current* is flowing. Alternating current has different laws.

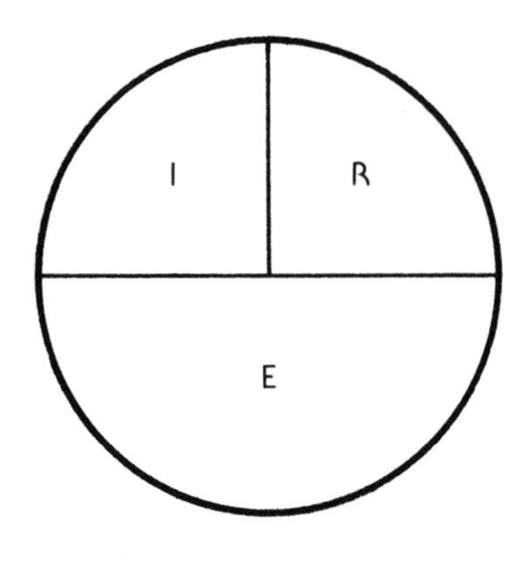

Figure 27B

We have already discussed the first version of Ohm's Law: The voltage is equal to the current times the resistance. The second version is: The current is equal to the voltage divided by the resistance. Increase the voltage and the current is increased if the resistance remains the same. The third version is: The resistance is equal to the voltage divided by the current. Again, the increased voltage means a decrease in resistance.

EXPERIMENT NO. 20: CHECKING OHM'S LAW

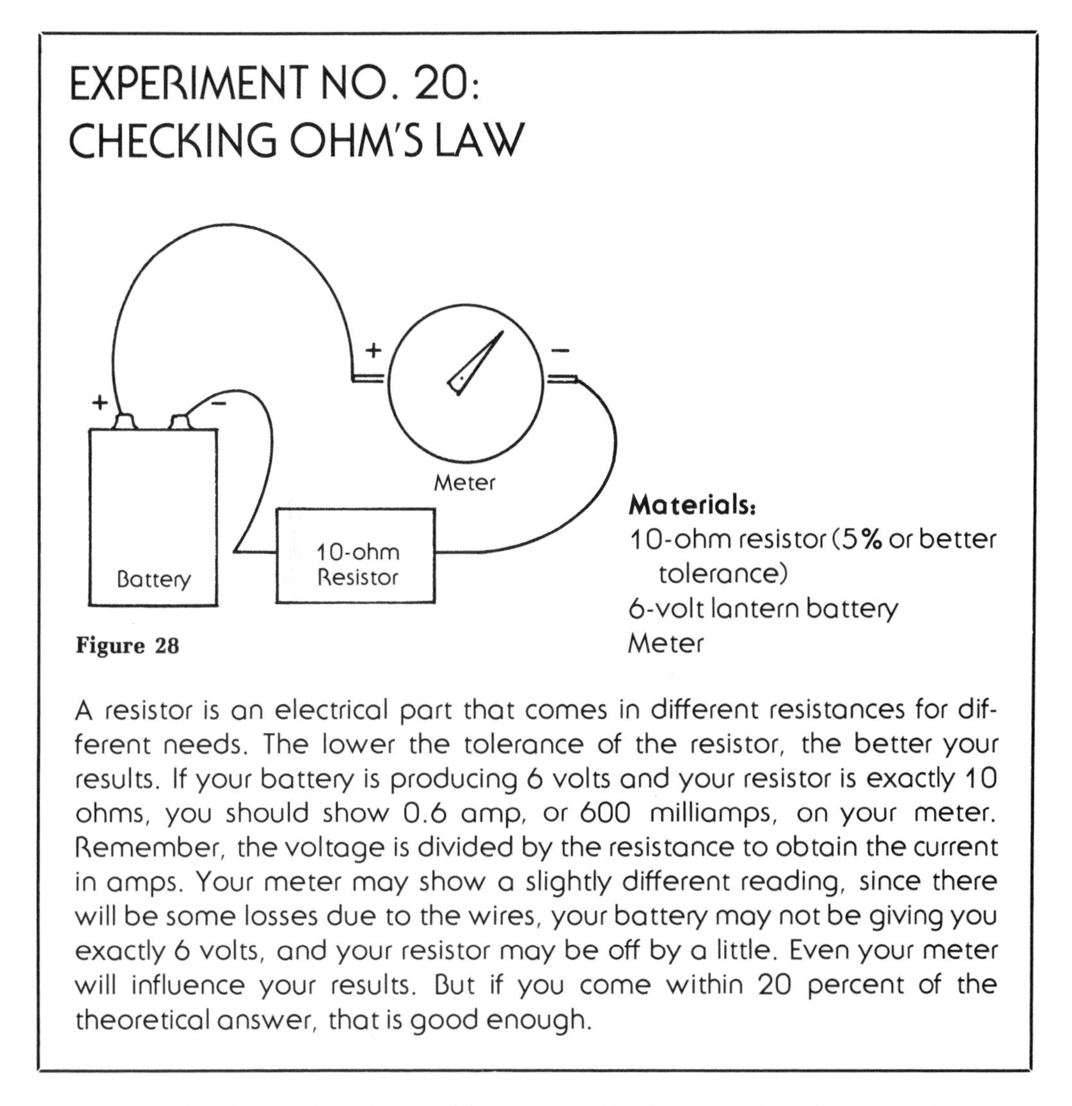

Figure 28

Materials:
10-ohm resistor (5% or better tolerance)
6-volt lantern battery
Meter

A resistor is an electrical part that comes in different resistances for different needs. The lower the tolerance of the resistor, the better your results. If your battery is producing 6 volts and your resistor is exactly 10 ohms, you should show 0.6 amp, or 600 milliamps, on your meter. Remember, the voltage is divided by the resistance to obtain the current in amps. Your meter may show a slightly different reading, since there will be some losses due to the wires, your battery may not be giving you exactly 6 volts, and your resistor may be off by a little. Even your meter will influence your results. But if you come within 20 percent of the theoretical answer, that is good enough.

You can now understand why George Ohm was so important. Although the relationship among volts, amperes, and resistances seems very simple, it took one man to figure it all out so that we can use it in all our circuits, no matter how simple or how complex.

Experiment No. 20 (Fig. 28) will help you understand Ohm's Law as it relates to an electrical circuit.

10

Batteries: Old and New

The Leyden jar permitted the storage of static electricity. But since it was not practical to move the jar, it had very limited applications.

The voltaic pile, as the first battery was called, employed copper and zinc in a strong acid solution. At least the pile could be moved about. Of course, it too had limitations. But it was to be the first step in the long road to portable electricity.

For years after Volta, an acid was the electrolyte in which a variety of metals were placed. The problems with such batteries were that the metals were usually expensive, the acid was highly corrosive, and the batteries were subject to almost instant discharge. Spilling the acid meant burnt clothing at the least and burnt skin at the worst.

(The words *batteries* and *cells* are often mistakenly interchanged. A battery consists of several cells linked together to produce a greater quantity of voltage or current. A modern car battery, for example, has six cells. Each cell produces about 2 volts, giving a total of 12 volts, the usual voltage in cars.)

Storing Electricity

Gaston Planté (1834–1899) was a French inventor whose idea was to make a battery that would store electricity. This was not a new idea. Secondary batteries, as the storage batteries are called, were known for years. They were of little value then, because the only way to charge them was to take the charge from another battery. There was no point to this exchange, and so the secondary battery remained a curiosity in the physics laboratory. Then along came the dynamo—a method of generating electricity—and a way of charging a battery became available. At once the old concept was dusted off and the secondary battery was developed.

Gaston Planté invented such a cell in 1860. He immersed lead plates in a solution of sulfuric acid. Once charged, such a battery could remain charged for days. The acid remained the danger it had always been, but instead of utilizing two expensive metals for the electrodes, he employed a relatively cheap substitute: lead.

The storage battery in today's vehicles is not dissimilar from Planté's invention.

Some time before, John F. Daniell (1790–1845), an English physicist, had devised a different type of cell. He was the first to use two liquid electrolytic solutions: copper sulfate and zinc sulfate. The electrodes were made of zinc and copper. The two liquids were not as corrosive as sulfuric acid.

Other inventors came along who improved on Daniell's cell. Different combinations of metals were employed. In some, the improvements consisted of rearranging the inner construction of the cell to produce more voltage, to give it a longer life, or because it could be made cheaper. None of these batteries were meant for popular use. They were intended for the laboratory to help in measurements and other scientific work.

Most of the cells never became popular—even in the lab—because they were often dangerous. Some sent up toxic fumes or had acids that could be handled only by someone wearing acid-proof clothing.

Dry Cells

Georges Leclanché (1839–1882) was a French engineer. He had a different idea about cells. In 1877 he discovered the way to use chemical action *inside* the cell to produce electricity.

He filled a cylinder with a weak solution of ammonium chloride (sal ammoniac). Into this cylinder he inserted a carbon rod and a zinc strip. This produced electricity, but the cell was nonrechargeable. Once the chemical acted upon the elements inside the cell, that was the end of it. It had to be thrown away.

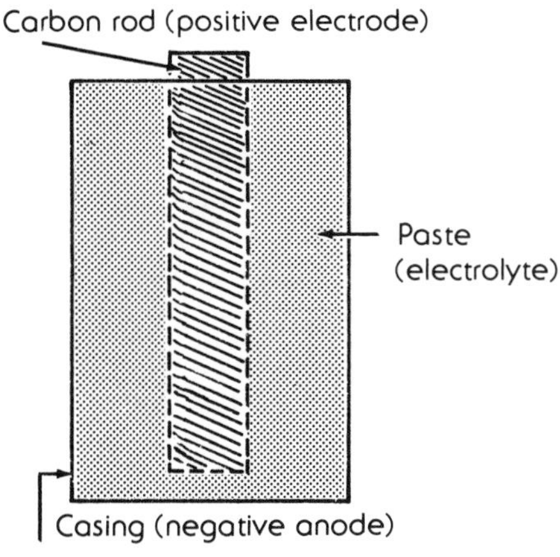

Figure 29. A typical dry cell.

Later he developed the first dry cell (see Fig. 29), which was not too different from those you use today for flashlights, toys, and portable radios.

The container and the negative electrode were made of zinc. The inside was filled with a paste made up of ground carbon, manganese dioxide, and sal ammoniac. A carbon rod going through its center was the positive electrode. To protect the cells from leaking—up to a point—a steel jacket was wrapped around the zinc container.

Many others were to come after Leclanché and improve on his work. He was the first, however, to conceive of the idea of making electricity really portable, and thus he paved the way for many modern inventions. Hearing aids, cameras, flash guns, electronic calculators—you can add to the list—none of these could work without the genius of the first man who came up with the idea.

Batteries of the Future

Today, continual work is being done to improve batteries. The goal is to invent a battery that will furnish more voltage without increasing its size and weight. Work is going on in many laboratories to develop a battery to run an electric car. (Did you know the word *laboratory* means "work shop"?) The batteries now used to run the completely electric car have to be recharged every 50 to 75 miles. Such batteries are being employed in a limited way in vans that deliver milk and groceries in rural areas in England and Germany. Each night the batteries are plugged into a special outlet in the garage to get the van ready for the next day.

When the right battery comes along, you will have a car that will be almost soundless. Imagine no more traffic noises! It will also be nonpolluting. There will be no need for a gear shift or transmission. To make the car go faster, you would merely furnish the motor more voltage. Maybe *you* might be the one to invent such a battery. Think how useful it would be to eliminate the smog that covers entire cities!

In the meantime, you might like to read about another inventor's contribution to the development of a battery. His name was Edison.

Edison's Batteries Fail

Thomas Alva Edison (1847–1931) was an American inventor with almost 150 patents on the battery alone. While his battery did not achieve great success, it was one more step toward a better battery.

Edison's battery was made of a nickel-oxide

positive electrode. The negative electrode was made of iron. The solution in which the two electrodes rested was a dilute solution of sodium hydroxide (potash). The battery was called an alkaline because the electrolyte was alkaline in its chemical composition. (It is not to be mistaken for the dry cells on the market today that are also called alkaline.)

It had many advantages over the lead-acid battery in that it was lighter and weighed less. Its basic flaw was that it did not work in cold weather, and it broke down very often.

However, Edison is best known not for his battery, but for his work on the transmission of light and power. He also improved the electric telegraph, the telephone, and the phonograph. We'll get back to him later when we discuss some of his many inventions.

Fuel Cells and Other Space Age Batteries

The concept of a fuel cell goes back almost 150 years. If it was so good, why wait so long to utilize it? The reason for its being put aside was that it was too expensive then to manufacture it. Now modern technology can produce it at a reasonable price.

The concept of the fuel cell is simple. It contains an electrolyte—an electrical conductor—between two electrodes. Sounds just like an ordinary dry cell, doesn't it? Now comes the difference. A fuel containing a lot of hydrogen in its chemical makeup, such as methane or propane gas, is exposed at one electrode. At the other electrode, a gas containing oxygen is employed. (Air is the obvious choice. It has a lot of oxygen, and it's free!)

Here's how the fuel cell operates: Due to the electron flow between the two gases from anode (positive pole) to cathode (negative pole), an electric current is generated. Hydrogen and oxygen combine. The result is water. This escapes from the cell as steam, since heat is produced by the union of the two gases. (This is the reverse of electrolysis. There you passed an electric current through water by means of two electrodes. You obtained hydrogen at one electrode and oxygen at the other.)

Fuel cells have already been used for moon landings. Closer to home, these cells are being tested for various types of vehicles.

The cells can be stacked to produce almost any amount of power. They are small, silent, and almost pollution-free. As the cells generate electricity, they also produce a valuable byproduct—heat. This could be used to heat nearby apartment houses.

A still newer battery is all plastic. Drs. Alan J. Heeger, Alan MacDiarmid, and Paul J. Nigrey at the University of Pennsylvania invented a plastic that behaves like a semiconductor—it can conduct electricity, and it can also store it.

The two electrodes are plastic. By making a chemical change in the plastic, one electrode becomes the anode and the other becomes the cathode—positive and negative.

No gases are formed so the battery is completely sealed. You can carry it in your pocket just as you would an ordinary dry cell. But there's another big difference: The experimental model I saw was about the size of the nail on my little finger. It weighed one-tenth of a dollar bill. It was so small an electric watch could have the battery buried in the bracelet. But in spite of its size, it ran a tiny fan.

The full scale model is expected to be able to power an electric car. While present car batteries produce only about 100 watts of power per pound of battery, the all-plastic battery will be able to generate 1,000 watts per pound. It can be recharged about 1,000 times before it is worn out.

There is still room for someone to come up with a still better or cheaper battery—how about it?

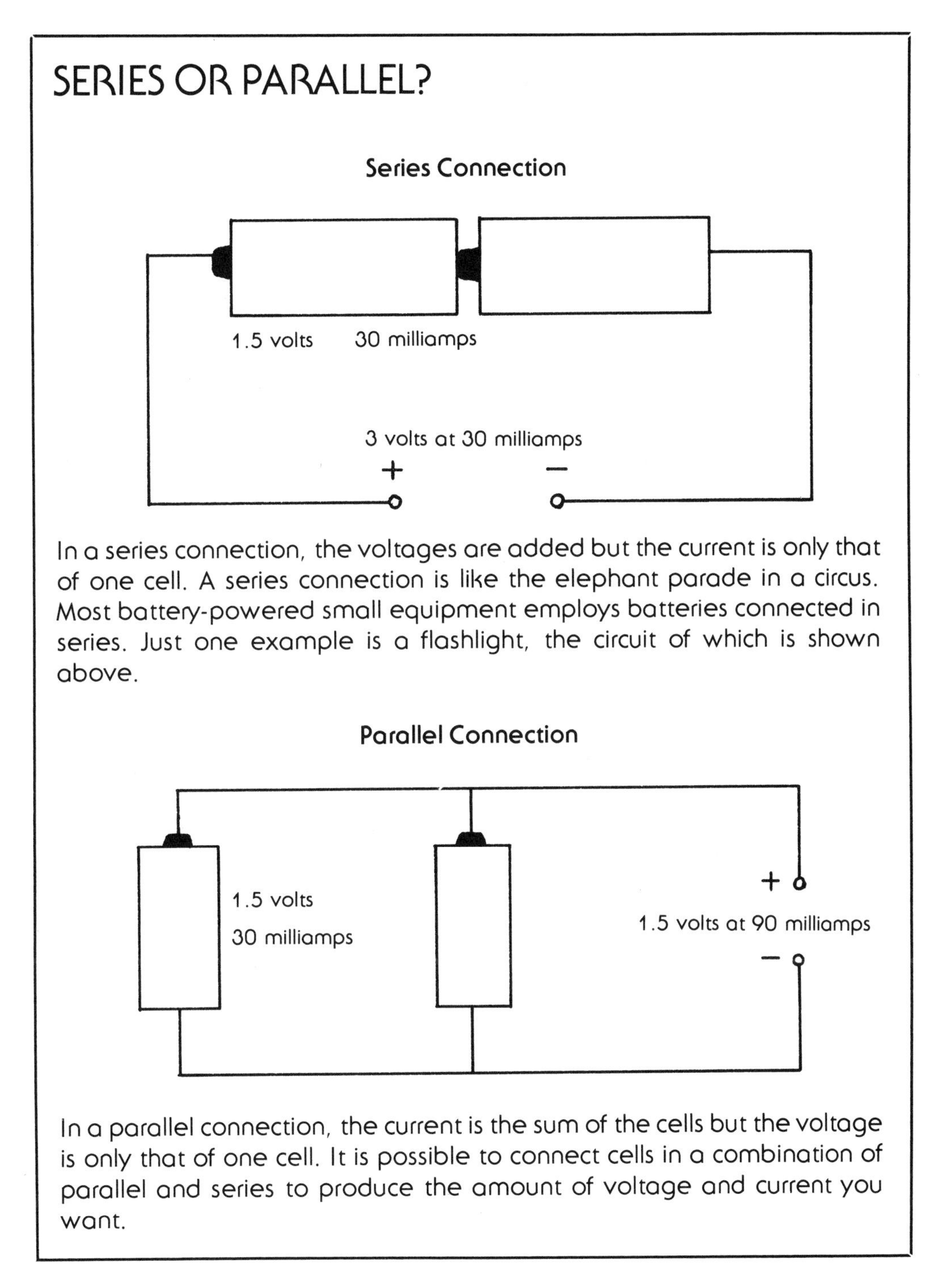

SERIES OR PARALLEL?

In a series connection, the voltages are added but the current is only that of one cell. A series connection is like the elephant parade in a circus. Most battery-powered small equipment employs batteries connected in series. Just one example is a flashlight, the circuit of which is shown above.

In a parallel connection, the current is the sum of the cells but the voltage is only that of one cell. It is possible to connect cells in a combination of parallel and series to produce the amount of voltage and current you want.

11

Battle of the Electricity Titans

Up until now, the progress of electricity was like going down one large highway. We traveled together through these pages very much in a straight line from one discovery to another, to another, to another. . . .

Suddenly we've come to a fork in the road. There are many branches leading from that fork. Down one road is the telephone, wireless telegraphy leading to radio is down another, and there are dozens of other paths. Some of these are not fully explored as yet.

One man traveled down several of these paths. He would go down one for a while, then work his way along another. He did this most of his lifetime. Sometimes he found a treasure at the end of a path, other times he found little or nothing. It would take others to penetrate deeper into the woods.

This man was one of several inventors who seemed to be able to advance electricity in several directions. His name was Edison.

The Incandescent Mr. Edison

Undeniably, America's most famous inventor is Thomas Alva Edison (1847–1931), the "Wizard of Menlo Park." During his lifetime, Edison received a total of 1093 patents. Of these, 141 were for batteries alone. He received 150 for improvements of the telegraph and 389 for electric power and light.

His fame rests on his invention of the incandescent lamp, which gives off light by burning a filament inside a glass bulb. The principle of the burning filament was known before Edison, but it was not considered practical. The light lasted only a few minutes and was very expensive to make.

It took Edison and his team (at one time, he employed as many as 3,000 helpers) a year of work, trying out hundreds of different materials for the filament, before they made a lamp that lasted more than a few minutes. He discovered along the way that the lamp would burn brighter and last longer if air was removed from the inside of the bulb. He even developed a better pump to obtain a better vacuum in the lamp. He perfected the incandescent lamp in 1879.

The lamp is the perfect example of a practical invention. It became so popular that Edison set up an entire factory just to make and market his lamps. Wherever in the world there is electricity, there are bound to be incandescent lamps burning.

As a youth, Edison had become a telegrapher after only four years of formal schooling. He saw where there was room for improvements on the electric telegraph. Even as a young man, he was full of ideas that would make money. He had no interest in scientific principles; he just wanted whatever he was inventing to work so he could sell it. As a result of his interest in the electric telegraph, he developed one on which several messages could be transmitted at the same time on the same wire.

Edison not only invented new things, but, as in the case of the telegraph, telephone, and the

Edison at his desk, puzzling over a problem. (Courtesy Consolidated Edison Company of New York)

phonograph, he tried to better them. For example, in 1888 he saw photographs taken in sequence so that they almost gave the appearance of movement. Immediately he thought of making a machine that would show these pictures so they seemed to move and that he could combine with his phonograph so that the actors and actresses would appear to be talking.

This idea was a failure. He could not get the two systems to work together so that the movements of the lips would match the sounds. Far worse, sometimes the woman would be speaking in a male voice or the actor suddenly had the high-pitched voice of a young girl. Not until 1927 was a film—*The Jazz Singer*—made in which the actor sang. So Edison anticipated the "talkies" by almost forty years.

Another of his dreams was bringing light into every home and every factory. In those days, houses were lit by gas jets and the streets of New York were illuminated by arc lights. Under his personal direction, the first central commercial incandescent electric generating station in the nation went into operation on September 4, 1882, in New York City. The station served about one square mile known as the First District. That first day he had only 52 customers who wanted his electricity. It was only the beginning of a network of utility stations offering electricity all across America. Today few homes cannot turn night into day by means of Edison's lamps.

Edison's generating station furnished direct current only. He believed very strongly that alternating current had no future because it was extremely dangerous.

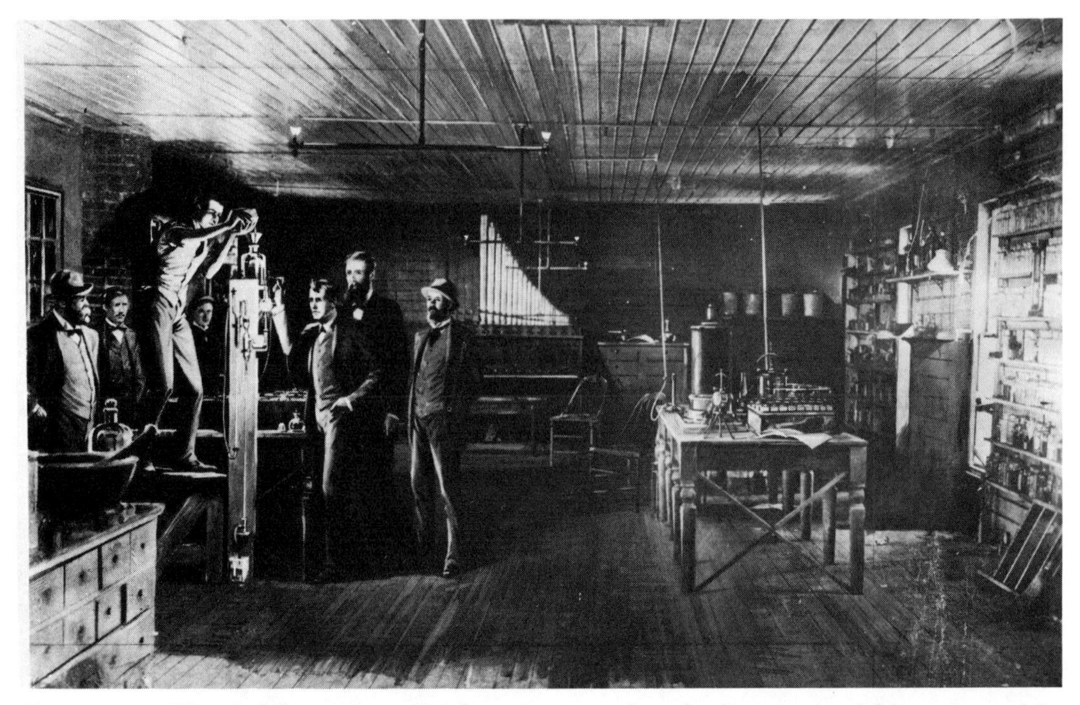

The scene is Edison's laboratory on October 19, 1879, when the first successful lamp burned for 40 continuous hours.

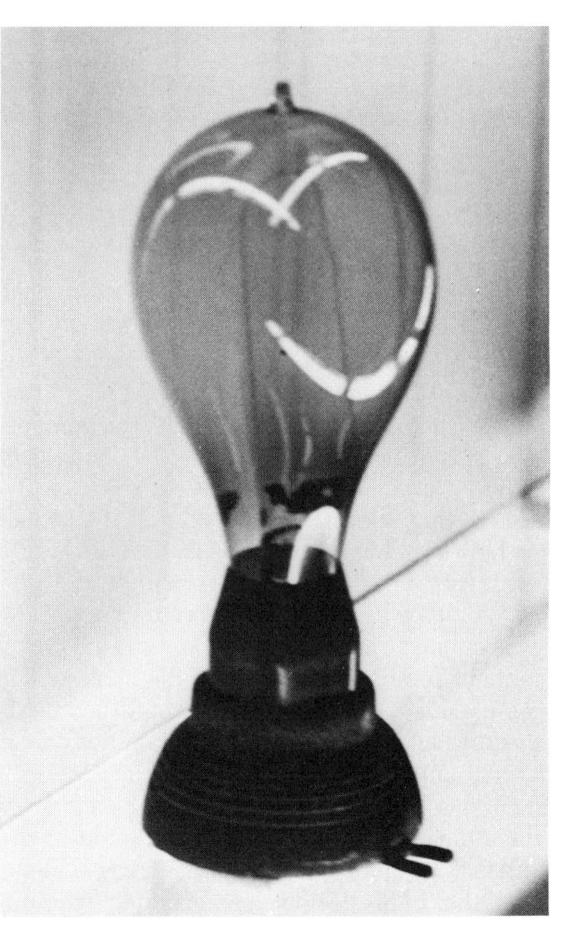

An early lamp made in 1886 producing the equivalent light of only 16 candles. Efficiency is low, only 1.4 lumens per watt. (Courtesy Con Edison Museum)

A modern sodium lamp with a high efficiency of 80 to 140 lumens per watt. This is 100 times that of the lamp of 1886. (Courtesy Con Edison Museum)

ARC LAMPS

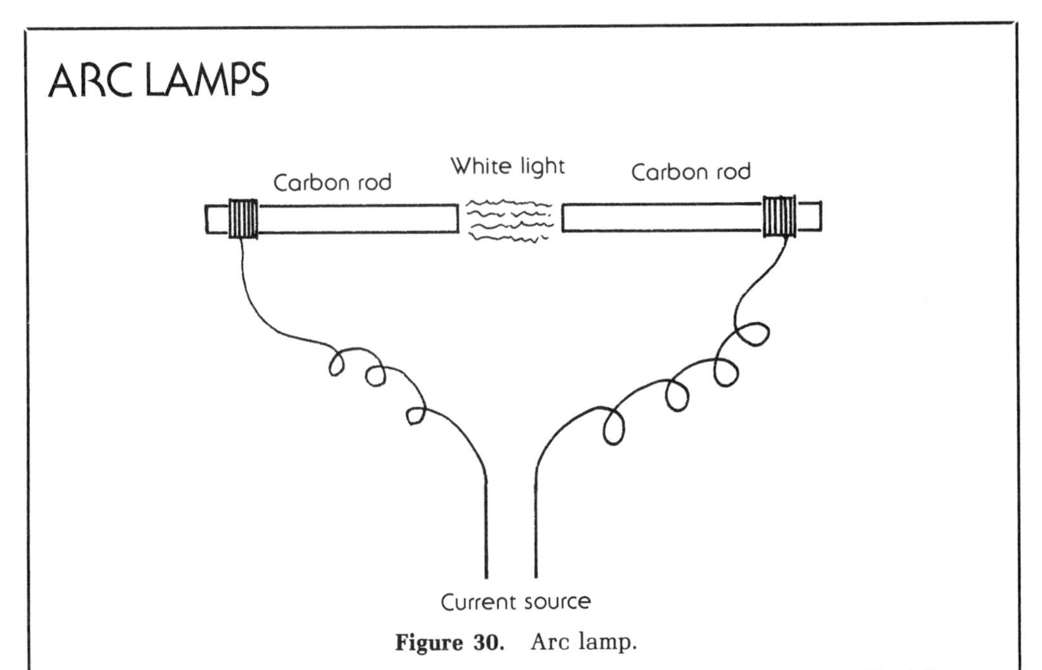

Figure 30. Arc lamp.

Streets in large cities were originally lit by illuminating gas (called "gas lamps"). Electric transmission made it possible to use the "new fluid" arc lamps (see Fig. 30). In an arc lamp, two carbon rods are connected to a source of voltage. A strong current is passed through the rods. The tips of the rods are brought together, a spark leaps up, and the light goes on due to the flow of electrons. Then the rods are separated until a tiny gap exists between them. An intense white light results.

There is a serious problem with such an arrangement. The light is intense until the rods are burned down and the gap becomes too great for the electrons to go from one rod to the other. At that moment the light goes out.

There were some mechanical devices that kept the rods automatically at the same distance. However, it is well known that the more complicated a scheme becomes, the more trouble develops.

Today rare gases such as argon, helium, and neon are utilized in the place of carbon rods when very bright lights are needed. The typical lamps seen on streets, highways, parking lots, and airports are designed to produce a great deal of light for their size and for the amount of current they consume.

You Can Be Sure If It's Westinghouse

One man in particular disagreed with him, taking the opposite stand—George Westinghouse (1846-1914), the founder of the company that bears his name. This began a bitter rivalry between AC and DC that continued for years.

Westinghouse founded his company based

The original tin-foil phonograph in 1877. (Courtesy Con Edison)

on his invention of compressed air brakes for railroad cars. It was an invention that made him rich. He jumped into the field of electricity by buying up patents. After a while, he was able to rival Edison. One of the inventions he bought was a transformer.

You remember that you did an experiment winding two coils around an iron core. You showed that you can *induce* current to flow through a coil that has no connection to the electrified coil (see Fig. 31). In your experiment, you wound the coils side by side. It was easier that way. However, commercial transformers are not made that way.

One coil is wound *over* the other. Both are heavily insulated so there is no possible shorting between them. By winding the coils over each other, you get maximum efficiency. The amount of induced current is partially dependent on the closeness of the two coils.

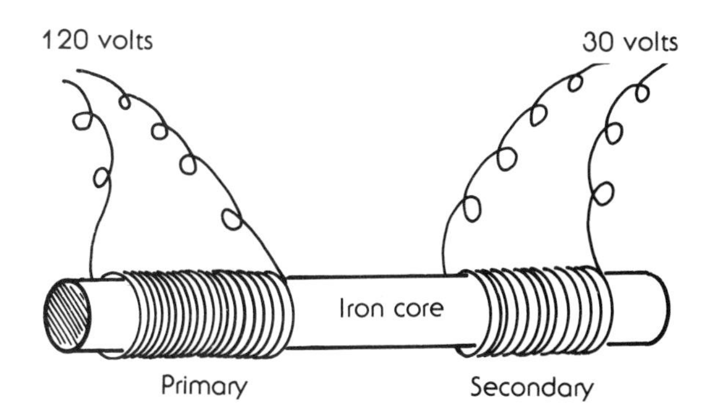

Figure 31. Transformer.

In Figure 31 above, the primary and the secondary coils are shown side by side for ease of illustration. But let us imagine that they are wound one over the other.

When the coils are close together, the *output* voltage from the secondary (nonelectrified) coil

Edison making final adjustments on the first modern picture projector. (Courtesy Con Edison)

is in the same proportion to the *input* voltage as the number of windings in the secondary coil as compared to the number of windings in the primary (electrified) coil.

In Figure 31, the primary is considered to have 100 windings. The secondary supposedly has only 25 windings. The proportion is 1 to 4. Therefore, the output voltage is in the same proportion of 1 to 4. Since the input voltage is 120 volts, the output voltage is one fourth of that, or 30 volts.

But you will find that a typical transformer will not be marked "30 volts." Instead it will be more likely to be marked "26.4 volts." The reason for that is that the manufacturer of the transformer knows there is bound to be some reduction in voltage due to resistance in the coils and heat loss. He also knows that input voltages vary. House voltage may go from 120 to 95 volts.

Page 60 shows a *step-down* transformer. The *output voltage is less than the input voltage.*

A *step-up* transformer is also possible. Turn the same transformer around. Make the secondary coil the primary. The proportional voltage will then be 1 to 4. The output voltage in that case would be 4 times 120 volts or 480 volts.

Readers may now be saying, "You are disagreeing with what you said before. You are creating electricity."

Electricity is *not* being created. You have to pay a price for stepping up the voltage. Double the voltage, and you reduce the current by half. Multiply the voltage by four, and you get only one fourth of the current that went into the primary.

As you realized, these transformers operate on alternating current. The newly invented transformer that Westinghouse bought had proved in experiments that AC was safe. The inventor, William Stanley, had improved some older inventions dealing with transformers. His transformer showed that you could transmit current along three quarters of a mile by first "stepping up" the original 500 volts from the generating station to 3,000 volts for transmission. Then at the other end he "stepped down" the voltage to 500 volts. The reason for the "stepping up" is that it is more efficient to transmit at higher voltages.

In most places, electricity is generated hundreds of miles from where it is needed. Cables are responsible for transmitting the electricity from point to point. Cables have resistance, and the transformers along the lines also cause losses due to their inner resistance. They also develop heat, which disappears into the surrounding air

To reduce cable losses, engineers found that if they doubled the transmission voltage, they reduced losses by almost three-quarters. That is another reason electricity is transmitted at very high voltages—in the United States generally at 345,000 volts.

At distribution centers, these voltages are dropped down in a series of steps until the electricity enters the house at 220 volts. The last step in the house is to step this down to 110 or 120 for household consumption.

None of this existed when Edison was fighting against AC transmission. He and Westinghouse were still battling for control over the goldmine of the century: the electrifying of America.

Electrifying America—not only bringing electricity to homes, but also selling generators and motors to factories—promised to be very big business. Companies were started up to take advantage of the demands for all of these new products that were appearing. Inventors were coming along with products based on the use of electricity that would make life brighter, easier, or just more enjoyable. Patents were being filed in Europe as well as in America. Electricity created the possibility of hundreds of new things, undreamed of just a few years before.

The outcome of the race between Edison and Westinghouse was anyone's guess until a new factor entered—an eccentric, brilliant inventor named Nikola Tesla.

Tesla's Amazing Inventions

Nikola Tesla (1856–1943) was born in Croatia, now a part of Yugoslavia. In spite of only two years of engineering college he visualized extraordinary, wondrously amazing inventions.

The areas of AC transmission, AC motors, and radio communications occupied him for most of his life. Above all, he delighted to work with high voltages at high frequencies. He had a rare ability to visualize in his mind the apparatus for an invention so clearly that the smallest details were there in front of him.

This was an advantage that worked against him as well. Usually an inventor or scientist keeps a notebook with details of his or her work—failures and successes—so that others who read the daily record can re-create just what was accomplished. But Tesla did not keep such a notebook, so many of his achievements are lost. Even many of the inventions that he demonstrated to scientific societies in America, England, and France cannot be duplicated. Thus, some of his inventions are doomed to be lost forever. Others may someday be rediscovered from hints he dropped or from vague ideas he set down once in a while. The trouble is that so many ideas poured from his mind that it is impossible to separate the actual concrete possibilities from his hoped-for dreams.

Tesla first went to work for Edison in one of the inventor's European companies, but after a time Edison brought him to America.

Tesla made the mistake of telling Edison about the induction (AC) motor that he wanted to develop. Edison was so personally involved with DC transmission and motors that he refused to recognize their advantages, and discouraged Tesla from going in that direction. But Tesla believed so strongly in his motor that he set up his own laboratory in New York City to work on it. In 1888 he announced his invention of the induction motor.

The original Pearl Street generating station built in 1882. (Courtesy Con Edison)

The site of the original generating station founded by Edison at 257 Pearl Street in New York City in 1882. (Courtesy Con Edison)

Transformers like this one vary greatly in size, but all work on the same basic principles.

At the same time, he was doing some spectacular demonstrations with high-energy, high-frequency electricity. George Westinghouse became interested in the young inventor after seeing those demonstrations, and hired Tesla to come to work for him. Westinghouse needed the AC motor to help sell AC transmission to American cities.

When the induction motor was offered, everyone agreed that it was superior to the DC motor. Everyone but Edison. But he had lost the battle. Even the company he started, Edison Electric Company, turned to AC. This company is now known as General Electric.

Today electric fans, air conditioners, refrigerators—every electric motor in today's appliances—are run on AC current. They are built on the principle of the rotating magnetic field originated by Tesla.

One of Tesla's dreams was to harness lightning as a source of free energy. If he could not harness it entirely, he hoped at least to duplicate the tremendous voltages in his own laboratory.

To enable himself to work with high voltages, Tesla invented the only machinery that bears his name—the Tesla coil. It consists of an air-core transformer with a spark gap and a capacitor. It is capable of generating extremely high (and dangerous) voltages at very high frequencies.

In 1890 Tesla discovered that by aiming a *low-power* Tesla coil at the human body, the invisbile rays could speed up the healing of sprains and torn muscles. He never took the time to patent this invention. Later it was patented by someone else. It is called diathermy or radiothermy.

There is also the possibility that Tesla was one of the first to see the potential of X rays. In a lecture, he described blurred photographic plates he had produced in his lab while he was experimenting with high-frequency voltages. As you may be aware, X rays are very short electromagnetic waves. Because they are so short, they are able to penetrate the body and show the shadow effect of bones.

One of the experiments that he carried out

AC INDUCTION MOTOR

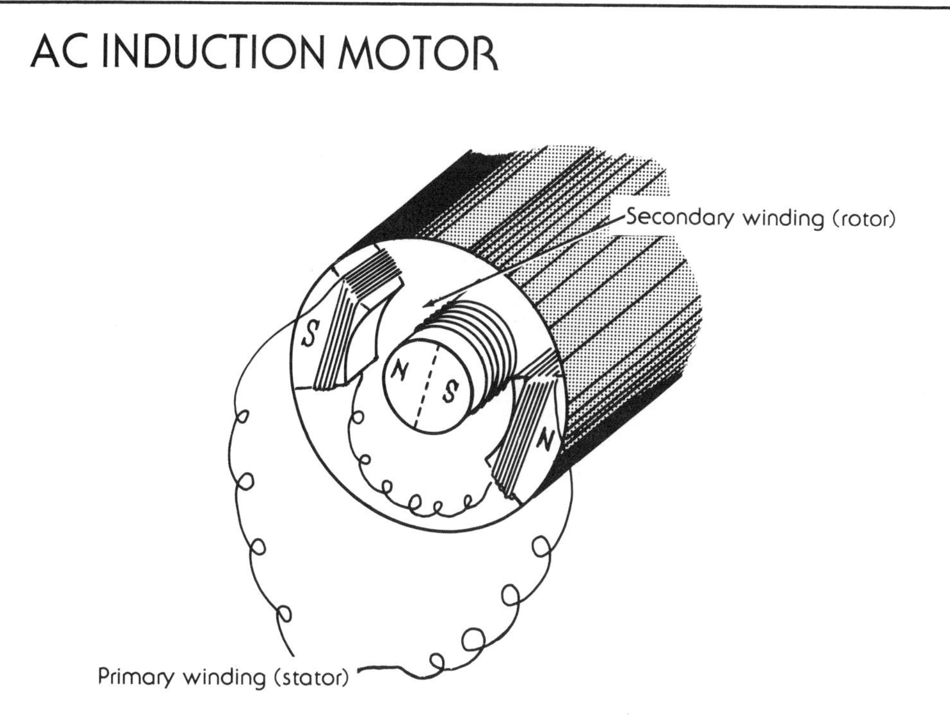

Figure 32. AC motor.

The induction motor—Tesla's invention—consists of a primary winding coiled around the opposite poles of an iron stationary ring. This is the **stator.** The secondary winding is around an iron core that is free to rotate on a shaft. It's called the **rotor.** Current flows only through the primary winding. North and south poles appear at the opposite sides of the stator. This action sets up a magnetic field (see Fig. 32).

As you recall from your experiment on a transformer, a current is induced in a coil that cuts across a magnetic field. The rotor's winding behaves the same way: the north pole of the rotor is attracted to the south pole of the stator, and vice versa. This creates an alternating current that changes direction 120 times a second in the U.S. Europe has a 50-hertz current (remember: 1 hertz = one cycle per second) that changes direction 100 times a second.

When the current changes direction, the poles of the stator are reversed. The rotor's poles also change as its south pole is repelled by the stator's old north pole that has suddenly become a south pole. The rotor makes a half turn. But as it does, the current has changed direction again.

If there were only two poles, as in the illustration, it would make the motor turn in fits and starts. To smooth out the movement, several pairs of poles are usually built into the stator. The principle remains the same, regardless of the number of poles.

The induction motor is only one of several types of motor capable of running on alternating current. But it is the type of motor you are most likely to meet in your home.

with a Tesla coil is worth describing. During a demonstration, he stood several feet away from his coil holding a Geissler tube. This tube is a forerunner of the modern fluorescent tube. When the switch to the coil was closed, the coil sparked and the tube he was holding in his hand, *unconnected in any way to the coil*, began to glow.

This was an early demonstration of radio waves. It showed that electricity at high frequencies could be transmitted without wires. Because of Tesla's uncanny ability to visualize an entire concept, he foresaw broadcasting in 1893, two years before Marconi was supposed to have discovered radio.

The fight over who deserved the credit for radio went on for a long time. It was not until 1943 that the Supreme Court of the United States stated that Tesla's patents were earlier than anyone else's in the field of radio transmission.

Tesla: An Inexhaustible Source of Ideas

At the time that Tesla was working for Westinghouse, there was no standard for the frequency of the transmission current. One, favored by many engineers, was 133 hertz (still called cycle per second or cps in those days). Tesla argued that it was not a practical standard. He wanted the standard to be set at 60 hertz. He eventually won.

Most of the North American continent has a standard of 60 hertz. This led the way for today's electric clocks, which keep time based on this frequency. Since the current changes direction 120 (twice 60) times a second, and since there are 60 seconds to the minute, we have an easy way to keep time.

The utility companies that furnish electricity to the homes furnish it at 120 volts. During periods of heavy demand, the voltage may go down to as low as 90 volts. But the frequency does not vary. Clocks are not affected by the small change in voltage. If there were a change in frequency, every type of motor that depends on frequency would be thrown off.

Tesla left Westinghouse to pursue his own experiments in the field of high-voltage, high-frequency electricity. With his machine he claimed he produced voltages of up to 20 million volts. He saw the earth as a huge magnet with an extremely powerful magnetic field around it. He believed that by applying the laws of electromagnetism, he could cut through the fields of force surrounding the earth and pluck electric power from the air around us. His goal was to provide an inexhaustible source of power to everyone on earth.

Recently, a scientist named Robert Golka found Tesla's diary for the year 1899 in the Nikola Tesla Museum in Beograd (Belgrade), Yugoslavia.

Golka used Tesla's notes to build a replica of his high-voltage machine. Golka sees high voltages as a means to achieve *nuclear fusion*. So far, he claims he has attained voltages of 20 million volts.

The form of atomic power we have now is *nuclear fission*—tearing atoms apart to create heat. Fusion, on the other hand, creates a sharp burst of intense heat by striking two atoms together—much like two cars colliding head-on. This form of nuclear energy would give the world a source of power without contaminating the air, the land, and the water. Fusion would produce no radioactive substances.

Success is still in the future. Who knows but that a concept undreamed of by Tesla will go even further than his most magnificent ideas?

Steinmetz: AC Genius

In 1889, soon after Tesla's arrival in America, another immigrant came who was destined to leave a lasting mark on American electrical transmission. He came from Germany, changing his name from Karl to Charles Steinmetz when he became naturalized. He was born in 1865 and died in 1923.

Steinmetz became a consulting engineer for

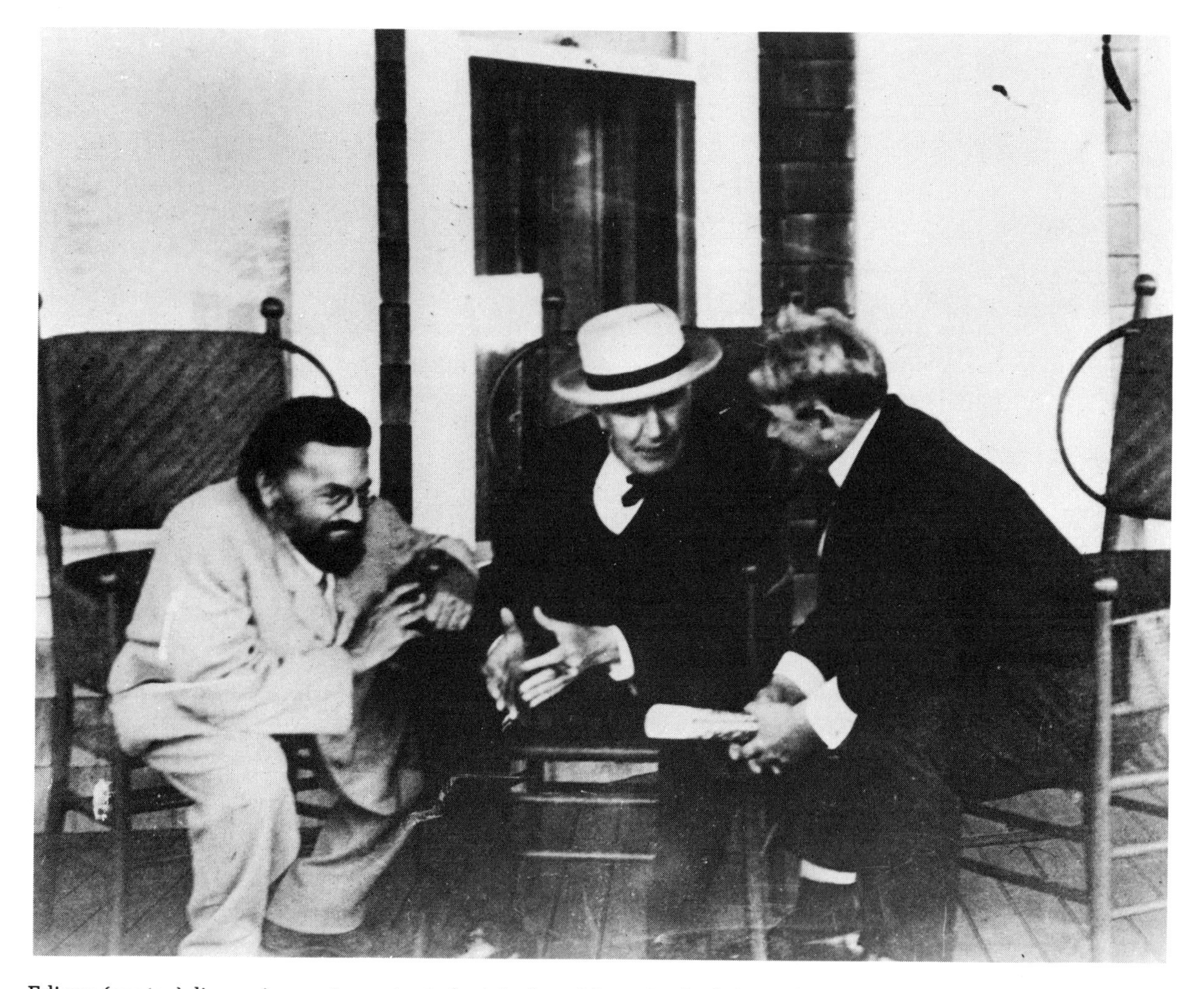

Edison *(center)* discussing an important electrical problem. At the left is Charles Steinmetz, another electrical "wizard." The man on the right is unidentified. (Courtesy Con Edison)

General Electric, Edison's old firm. During his nine years there, he spent time on the theories and calculations of AC. Much of his investigation went into the mysteries of electrical phenomena.

General Electric could not use the transmission system developed by Tesla because it belonged to Westinghouse. So Steinmetz produced a different version. During his lifetime he patented improvements on AC generators and motors.

His work was not flashy. Since it did not lend itself to demonstrations with lightning flashes, Steinmetz is not known to the general public. His contributions are important, however. They provided one more step in the progress of the science that went from magnetism to electricity.

12

Motors and Generators: Electrical Producers and Users

A Close Relationship

There is actually little difference between the operations of a DC motor and a DC generator. The motor converts electrical energy into mechanical energy, while the generator, or dynamo, does exactly the reverse. It takes mechanical energy and converts it into electrical energy.

To activate a motor, you feed it electricity. That makes it turn a shaft, and that produces work of one kind or another. To activate a generator, you make the core of the generator spin. It is this core or armature that produces electricity.

If you look at the photograph on the opposite page, you see what appears to be the same mechanism photographed twice. Actually they are both from the same kit sold by Radio Shack stores.

The one on the left is made to operate as a simple DC motor. The motor turns when you pass current through the armature. This turns the shaft that rotates a belt that is connected to the shaft on the generator. (The mechanism on the right.) As the armature of the generator begins to spin, it generates electricity. This is drawn from the two posts that are wired to a meter.

The DC motor was discovered because it is so much like a generator. In 1873 a Belgian engineer named Z. T. Gramme placed two DC generators side by side at a trade show. Supposedly, a workman connected one of the generators, which was steam-driven, to the second generator. The latter started spinning—behaving like a motor.

For the first time there was available a small powerful source of power that could run for days with little or no attention. Think of all the appliances right in the home that benefit from such a motor: electric fans, vacuum cleaners, air conditioners, and don't forget the tiny motors that cause the tape to run in portable tape recorders.

What Makes DC Motors Tick?

Let's look at a DC motor to see what makes it operate. Figure 33 shows a simplified version of the motor. The armature revolves between two magnets, called pole pieces. The brushes, made of brass or carbon, are stationary but touch the *commutator*. This is on the same shaft as the armature and revolves with it. The commutator, in its simplest form, consists of two pieces, insulated from each other. One brush is connected to a positive source of electricity, the other to a negative source. When the current is switched on, the armature becomes a

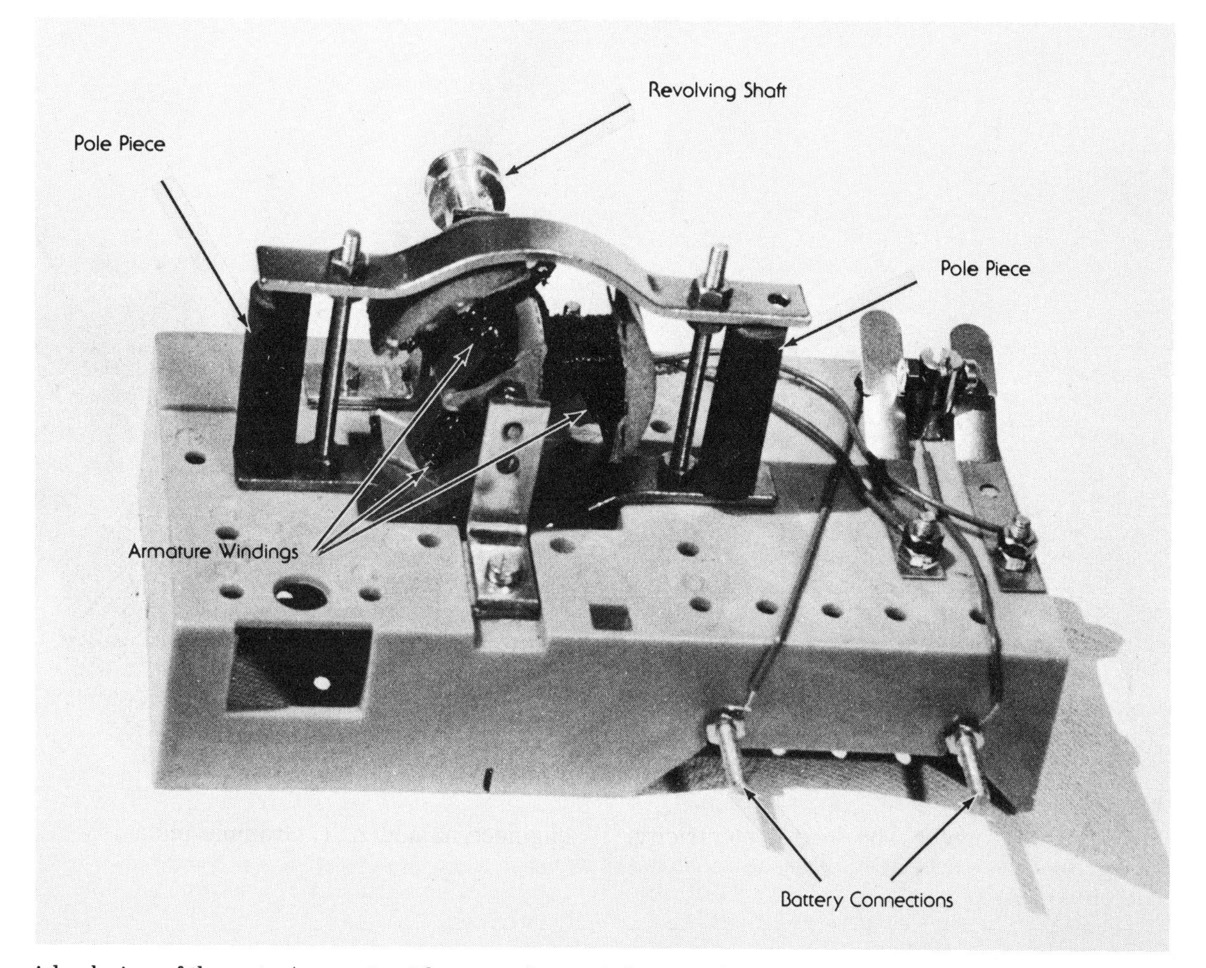

A back view of the motor/generator. There are three windings on the armature. A more powerful motor would have six and each would be larger. A commercial motor would use electromagnets rather than the permanent magnets that form the field poles in this tiny motor. (Courtesy Radio Shack/Tandy Corp.)

magnet due to the current flowing through the coil. As a magnet, it, too, has a north and a south pole.

Like poles repel each other, as you know, and unlike poles attract each other. The *north pole* of the armature is attracted to the *south pole* of the pole piece.

The revolving armature causes the commutator to revolve also. The half of the commutator that touched the positive brush has turned around so that half is in contact with the negative brush. The poles of the armature are reversed. The armature makes another half turn. This goes on and on as long as there is a source of power.

The little motor in the photograph has three windings for the armature. The commutator is also made in three separate parts. This "kicks" the armature around more often with less chance of stalling.

Depending upon what force they must possess, large-scale motors will have bigger or smaller windings. But they will generally have six windings to completely eliminate stalling. (If a motor stalls, the wires around the armature become heated and may even melt.)

DC motors were the only motors available for many years.

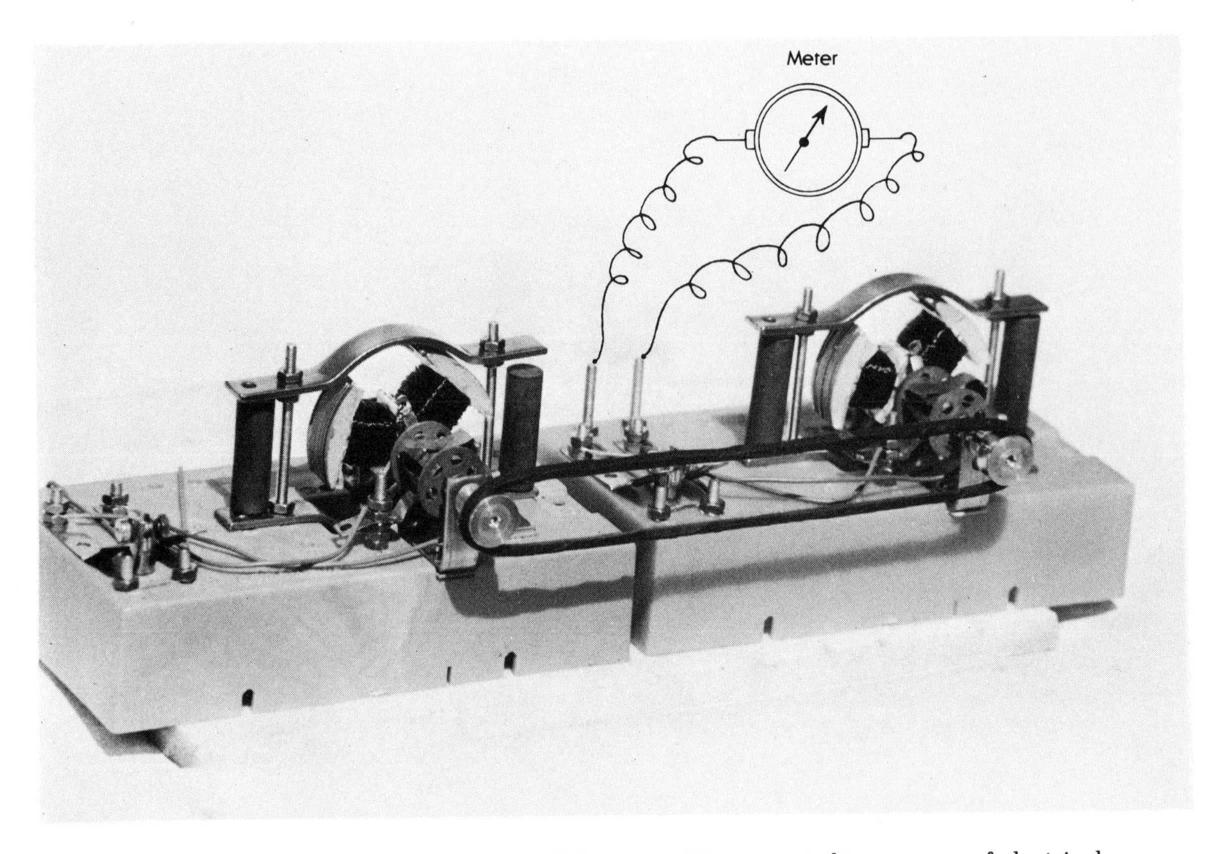

The two machines are identical. The one on the left (a motor) is connected to a source of electrical power. As it spins it turns the shaft connected to a belt. The right-hand machine (a generator) revolves and produces a tiny amount of electricity. (Courtesy Radio Shack/Tandy Corp.)

What Makes DC Generators Tick?

How different is a DC generator from a DC motor? There is little physical difference. Only the results are turned around.

Let's analyze the DC generator (see Fig. 34).

Faraday showed that when a conductor—for example, a coil of wire—cuts across a magnetic field, an electromotive force is generated and flows through the conductor.

In a DC generator, a coil of wire—the armature—is made to revolve between two magnets. This action cuts across the magnetic field of force. A current is generated that flows to the brushes. They are connected to a motor or to a battery to store for later use.

The amount of current generated depends on several factors: the number of revolutions, the size of the coil, and the strength of the magnets.

The small-scale motor in the photograph on page 66 is easily transformed into a generator. Disconnect it from the source of power. Instead connect the brushes to a meter, or to a tiny light bulb. Wrap a string around the pulley and pull it very quickly, as if you were starting a top. The meter will indicate current flow. If you wind the string in the opposite direction and pull it, the armature will also reverse. The poles of the commutators are also reversed and the positive side of the circuit is now negative.

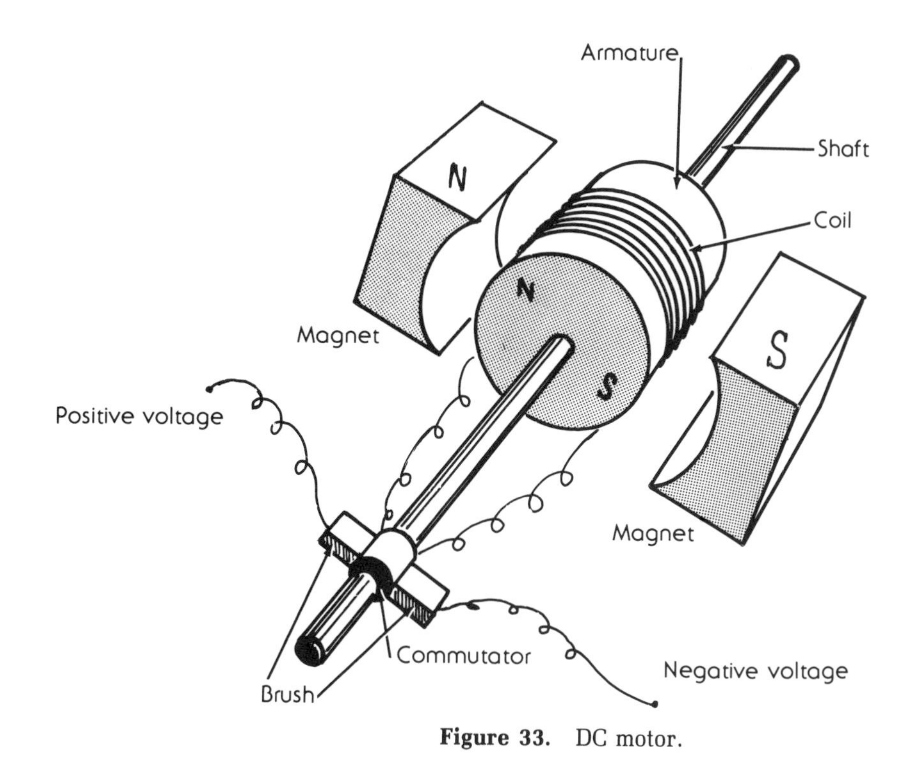

Figure 33. DC motor.

Perpetual Motion

Since you are generating electricity, why not use that electricity to run the motor? The generator would furnish power to the motor that would run the generator. If you could do that, you would accomplish what inventors have been trying to do for centuries—invent perpetual motion.

What happens in this case is that a great percentage of the current is lost due to the resistance it meets going through the coils. No matter how carefully a piece of machinery is designed, there will always be some friction losses.

Let's say that you use 6 volts at 300 milliamps to run your motor.

$P = E \times I$ (power ± voltage × current)

P (watts) = 6 (volts) × 0.003 (amps)
= 0.018 watt (or 18 milliwatts)

But at best the generator will not produce more than 40 percent of the input power needed to run the motor. It's barely possible that you will obtain more than 0.0009 amp (0.9 milliamp)—less than 30 percent efficiency. (Efficiency is the ratio between input and output.) Your motor/generator combination has lost almost 70 percent of the input power. So the dream of perpetual motion is still very much a dream.

In fact, the largest utility plants that generate electricity do not obtain more than 40 percent efficiency.

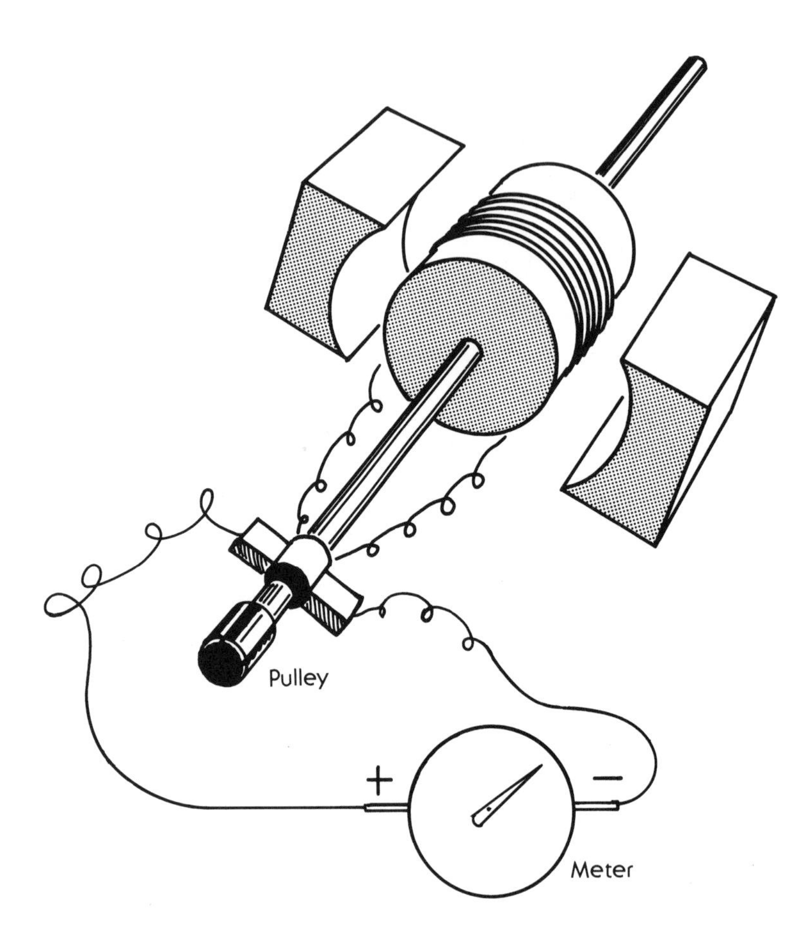

Figure 34. DC generator.

13

Pioneers in Communication

Alexander Graham Bell (1847–1922) was a Scotsman who emigrated to America. While we know him as the inventor of the telephone, he was equally famous as a pioneer in the education of the deaf. For this reason, he was interested in Edison's phonograph for recording and reproducing the voice.

During related experiments with the telegraph, he found that a device containing vibrating reeds could generate sound. From that he got the idea of making a steel disk vibrate by speaking against it. The disk was placed near an electromagnet. The disk made a pattern of vibrations on the current passing through the electromagnet. The pattern could be duplicated so a listener at the other end could hear the vibrations.

It was not a success until Bell hit upon the idea of using a *modulating* current to parallel the waveforms of the voice. To modulate a signal is to vary some characteristic of one wave in accordance with another. (The same idea was to be used by the inventors of radio.) The current must match the varying volume to make the voice seem lifelike.

Later a tiny cup filled with ground-up carbon was added. It is still used in today's telephones. A current passes through the carbon granules in varying levels. This modulates the current and allows it to reconstruct the voice at the receiver.

To honor Bell for his achievement, a unit of sound level is called a *bel*. Sound levels are measured in tenths of a bel, or decibels. The abbreviation for *decibel* is *dB*.

At about the same time, Hermann von Helmholtz (1821–1894) was working in Germany in a different area of electricity. He, too, developed measurements and evolved laws about the behavior of electricity and magnetism. His development of the mathematics dealing with resonance in an electrical circuit was years ahead of its time. The concept of tuning two circuits so they answer to the same frequency led years later to the invention of the radio.

One of the most versatile scientists of the 19th century, Von Helmholtz did considerable work in the science of acoustics. He showed that the quality and naturalness of the voice over the telephone was not due to a single tone. To sound natural and to be understood, the voice had to be transmitted as a combination of tones. These combinations are called *harmonics*. Strike a note on a piano or on a guitar. If you listen carefully, you will hear not only the note struck, but also a tone an octave above and one an octave above that. This makes the note sound rich and full.

Later, we'll get back to resonant circuits and how they are used in radio today.

Heinrich Hertz (1857–1894) was associated with Helmholtz, working as his assistant. By continuing the theories of Faraday and Maxwell, he laid the foundations for the future development of radio, telephone, telegraph, and even television.

Hertz did a great deal of work on the theory of waves. After studying the mathematical equations of Maxwell, he determined that there were electromagnetic waves all around us in space.

One of the questions that scientists in the 19th century were asking themselves was: When forces act upon each other at a distance, is it due to some medium between them, or in spite of it? Hertz proved that there was such a medium and that forces acted *because* of that medium. This is the wave theory that is accepted today. Because of his work in that area, they are called Hertzian Waves.

His proof rested on a simple experiment. Hertz made a rectangle out of heavy copper wire with a short gap between the two ends. The wire was connected to a circuit through which an induction coil was discharged. A spark jumped the gap. Several feet away, Hertz had a similar circuit, and he was able to get the spark to jump the gap in this second circuit with *no connection between the two circuits.* The second spark was a little weaker, but he showed that the two circuits, *if they were identical,* resonated. They vibrated in unison. This arrangement is called a *tuned circuit.* The tuned circuit allowed Hertz to detect electromagnetic waves at a distance from their origin. With his equipment, he could cause a spark to appear at the gap of the second circuit at a distance of up to 60 feet (20 meters). This was the beginning of the wireless telegraph and you may have guessed it—radio and TV. Without the tuned circuit, none of this equipment could work.

The experiment proved without a doubt that electrical waves traveled through the air, without these waves, radio and television could not work, nor could satellites beam messages back to earth. Hertz also showed that electricity could be transmitted by electromagnetic waves.

This scientist's name is carried on by the term we use to measure frequency: the *hertz.*

Guglielmo Marconi (1874–1937) was an Italian inventor best known for his contributions to radio. He was only 20 years old when he began his experiments to transmit a spark from one point to another without any connecting wires. He was employing electromagnetic waves to transmit a signal without any connection. He used equipment similar to Hertz's, but his apparatus was very much of an improvement over the spark and induction that Hertz employed.

At first, Marconi was able to transmit the spark only a few feet. In 1895 he was able to transmit from shore to a ship at sea. By 1896 he took out his first patent for wireless telegraphy, based on Hertz's discoveries but using much longer wavelengths. Marconi's first equipment consisted of a receiver with a ground wire connected into the earth and a high aerial.

He continued his experiments to extend the range of his wireless. In 1901 he sent the first transatlantic signals from Cornwall, the Southwestern tip of England, to St. John's, Newfoundland, some 2,000 miles (3,500 kilometers) away.

In 1909 he shared the Nobel prize in physics.

Some years before, a curious phenomenon had been noticed. When a signal, such as a conversation, was transmitted over a telephone line, a wire parallel to the first would pick up the conversation. In sensitive electronic circuits, great care must be taken by the design engineer to avoid having parallel wires for that reason.

The strange behavior did not go unnoticed by Edison. He filed it away. Then years later he remembered this oddity when he wanted to have a telegraph on a moving train.

He strung an aerial over several railroad cars. As long as the train ran parallel to the telegraph wires strung along the track, it was able to pick up the signals.

Edison received a patent on this idea, but this form of "wireless" telegraphy was not a success. Too often the wires were not parallel to each other, resulting in a lost or very weak signal. In spite of Edison's failure, when Marconi formed an American company to exploit his new wireless, he found he was infringing on Edison's patent.

Edison was slightly deaf. He could not visualize any future in what was to become radio. So he sold his rights for a few thousand dollars to what was destined to become a multi-billion-dollar industry.

During the first World War, Marconi experimented with short waves for a beam system for military signaling. When peace arrived, he continued his work with short waves, contributing to the refinement of long-distance communications.

14

Radio: AM & FM

A Smooth Beginning

It has already been stated that electricity's highway reached a fork and there were several roads to explore. One such road was radio. It progressed from the simple experiments beginning with wireless telegraphy to our present ability to reach almost every part of the world by means of satellites.

Exactly what were the first steps in the field of wireless communications?

If there is one specific point, it was Heinrich Hertz's discovery that electromagnetic waves existed in space. He showed that we are living in an ocean of these waves. They are all around us. We cannot hear them. We cannot see them. But in 1887 Hertz proved they existed in spite of their invisibility through his experiments with an induction coil and a spark gap.

Edison's attempt to transmit telegraphic messages without wires from a moving train to parallel wires was still one more step. By a coincidence, he tried out his system at just about the same time as Hertz was doing his experiments.

Guglielmo Marconi, using practically the same equipment as Hertz, was able to make his unit more powerful by means of a component then called a *coherer*. We now call this a *detector*. The French inventor of the detector found that when high-frequency alternating current was passed through metal filings, it increased their ability to conduct a current.

By means of the detector and an aerial, Marconi sent a message without wires for a distance of 1 mile (1.6 kilometers). A little before this event, Tesla also demonstrated the existence of electromagnetic waves by causing a Geissler tube to light up without connecting wires.

Marconi was a very practical man and immediately filed for a patent for the wireless. The basis for radio was, and still is, a *tuned circuit*. The transmitter and the receiver must be tuned to the same frequency. When you tune in a radio station, you are tuning the radio circuit to the same frequency as the transmitting station whose music you want to hear.

Marconi's invention, the wireless telegraph, became an actual lifesaver for ships at sea. Previously they could communicate with each other or with a shore station only when they were within sight. Mirrors (heliographs) and signal flags were the only means of signaling. With Marconi's invention, rescue vessels were able to arrive quickly after receiving an SOS. Laws were passed requiring every ship to have a wireless set.

Problems, Solutions, and Innovations

But Marconi's invention had one serious drawback. Communications could be carried on only by means of dots and dashes. A voice could not be heard because the spark gap transmitter was not adequate for the complex waveforms of the human voice. A "carrier" wave was needed, and it had to be in the same radio frequency range as the voice. The voice rode "piggy-back" on that carrier.

An American, R. A. Fessenden, tried to find the solution to this problem. He developed the idea of a high-frequency alternator operating at 50,000 hertz. This wave was at a steady voltage. A voice could ride that wave with a frequency of about 2,000 hertz. In 1906 an alternator following Fessenden's idea was built by General Electric.

At about the same time, improvements were being made on vacuum tubes. With these two elements—the carrier wave and the improved tube—radio was on its way. It was helped by a new, very powerful transmitter in New Brunswick, New Jersey.

Another American inventor gave one more push to radio. His name is E. H. Armstrong (1890–1954). He is credited with inventing the heterodyne and superheterodyne circuits. By feeding part of the output of a tube back to its input—this means from the grid of the tube to its plate—the circuit would oscillate. This is another way of saying that it would produce a steady tone. A high-frequency current would result and be used as the carrier wave.

Armstrong was also the inventor of FM (frequency modulation) radio. He fought a long battle against the networks who "borrowed" his circuits without paying any royalties. Before the courts decided in his favor after many years, he died. His invention lives on because FM produces a higher quality signal than does AM. It is not affected by natural static, lightning, or man-made static such as electrical generators or auto engines. Thanks to Armstrong, we have stereo FM and FM audio on TV.

But we're jumping ahead of the story about radio. Let's go back to Edison.

When Edison was trying to develop a successful light bulb (see Fig. 35A), he observed a curious phenomenon. The glass around the bulb blackened near the positive terminal of the filament. Looking for an explanation, he inserted a metal plate inside the airless tube. He placed the plate near the filament. Then he connected a meter to the plate and the filament.

Once the filament was heated, a small current was generated. He reversed the process. He caused current to pass through the filament and the plate. He found that the DC current could flow in one direction but not in the other. Edison saw no value in this and ignored it. What he wanted was a light bulb that would work.

Others saw the value of the "Edison Effect." An Englishman, J. A. Fleming (1849–1945), carried out further experiments with the Edison bulb. In 1904 he patented his own version of the bulb in which he added a metal cylinder that surrounded part of the filament. He hoped that his invention would serve as a detector. In this he was disappointed. It did work, however, as a *rectifier*. Rectification converts alternating current into a form that is almost direct current for practical purposes.

This became the Fleming tube, or valve as it is called by the British. The tube is actually a diode in that it contains two electrodes (see Fig. 35B). One electrode is the anode and the other is the cathode. The current flows from the cathode to the anode. Due to this advance, Fleming is considered one of the pioneers of radio.

An American was next on the scene. Lee DeForest (1873–1961) is called the Father of Radio. He patented 300 inventions in wireless telegraphy, radio telephones, and even "talkies" with the sound track on the film itself. He is best known for the improvement he made on the vacuum tube by adding a third electrode (see Fig. 35C).

The *triode*, as this tube was called, has an anode, a control electrode called a control grid, and one more electrode. The plate is positively charged and the cathode is negatively charged. The control grid has holes in it to allow the electrons to pass through. The number of electrons that pass through is dependent on the electrode charge.

This tube was to be employed not only as a detector, but as an amplifier. It would accept a small, or low-level, signal and amplify it to the point where it could be heard over a loudspeaker. Until then headphones had to be used because the signal was not powerful enough to vibrate the cone of the loudspeaker and cause a sound to be heard.

More improvements were added to the vacuum tube to achieve better results. Even-

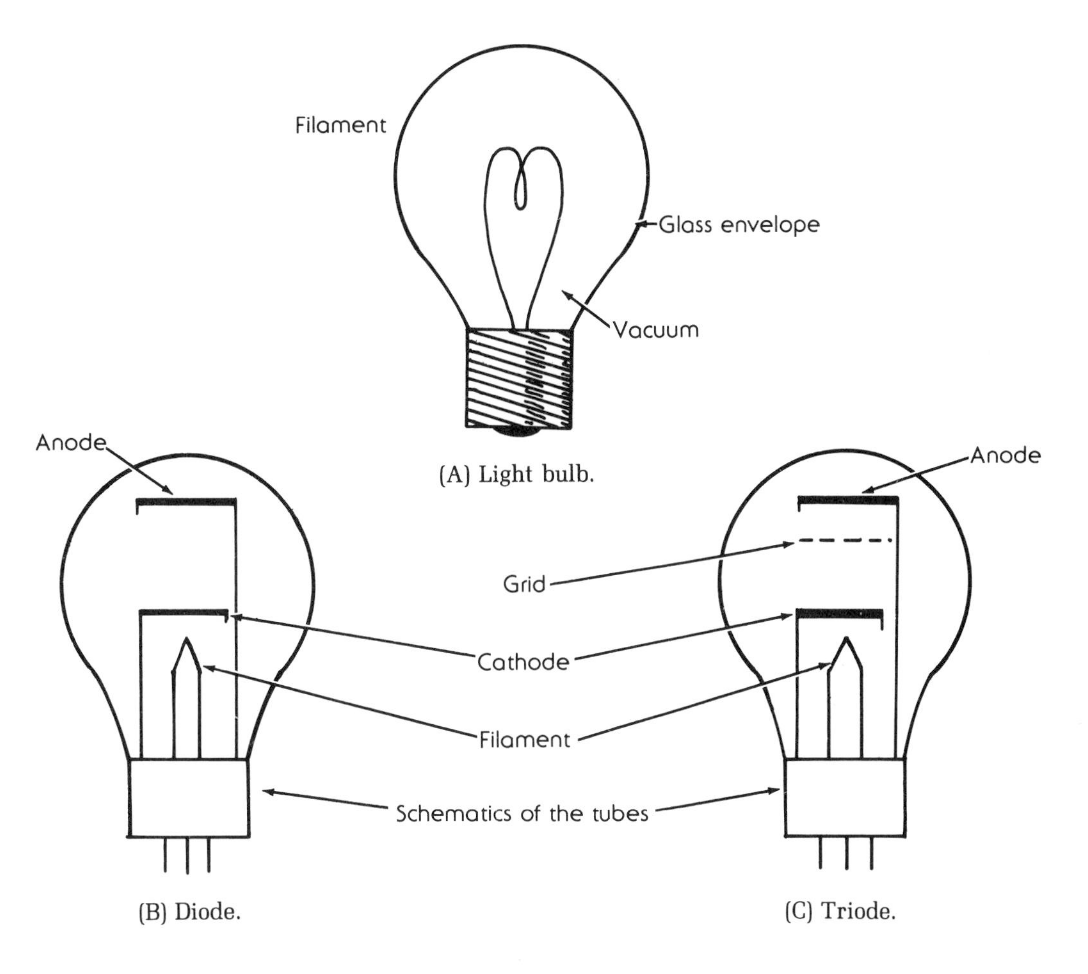

(A) Light bulb.

(B) Diode.

(C) Triode.

Figure 35

tually tubes had five elements and so were called *pentodes* (*pent* means "five").

There seemed to be no room for more advancements. And then came a discovery that was to make possible smaller electronic circuits requiring less voltage and lasting longer. Within a few years of this discovery, tubes were employed only for special purposes. The name of that revolutionary element was the *transistor*, which we'll learn more about in the next chapter.

15

Electronics: From Semiconductors to Solar Cells

That branch of science we call electronics began with the invention of the vacuum tube. Tubes were improved over the years until they reached a point where no more improvements could be made. Yet they had a serious disadvantage for the communications industry. Tubes could not detect extremely short waves, those we call microwaves.

Since tubes offered no hope in solving that problem, the communications people turned to Bell Telephone Laboratories. Bell Labs does research and development particularly for the telephone industry. Their scientists were asked, "What can you find that will transmit microwaves?"

As far back as 1874, a German physicist noticed that there existed a strange flow of current in certain minerals. These were grouped under the name of *semiconductors*. The word is used to indicate substance that can conduct a current better than an insulator, but not as well as a true conductor.

Nothing was done at the time about this 1874 discovery. Marconi used a "cat's whisker" detector for his high-frequency waves. It was also used in the very early radios. A piece of galena (lead sulfide) would be touched with a sharply pointed flexible wire. The wire was so thin that it was called a "cat's whisker." These semiconductors could detect radio waves received by an antenna. Little power was used by this radio, so headphones had to be clamped to the ears. You would move the fine wire around the surface of the crystal until you could hear a tiny, tinny voice crackling in your ears.

This was not what the telephone companies were looking for. So the scientists began many experiments with semiconductors, one of which was silicon.

The purest silicon was melted down and made into ingots or large bars. They found that there were two types of silicon: a positive or *p* type and a negative or *n* type.

The difference between the two types depended on impurities so small that they could not be detected by chemical analysis. They were classified depending on which way they favored the direction of the current.

Then, in 1947, the scientists found something very strange. Not all silicon was *p* or *n*. Some was *p* at one end of the bar but *n* at the other end. But the strangest part was the behavior at the junction where the two types met. The dark opaque material transformed light into electricity! Shining a flashlight on that area caused the needle on an ammeter to indicate current flow.

Another peculiarity of this crystal was that it acted as a *rectifier*. A rectifier converts (or rectifies) AC—that's the current that is continually changing direction—to DC. The latter flows in only one direction.

Early radios needed rectifiers because they could operate only on DC, so a special tube had been made that rectified the AC to DC. But suddenly here was something very small that could do the same thing without wasting a lot of voltage.

No one guessed at the double importance of the *p-n* junction at that time. The junction not only held the key to the transistor, but it also was the key to the Bell solar cell. Without solar cells, it would be impossible to recharge the batteries of space satellites by converting sunlight into electricity.

Several years were to pass before these would be put to practical use. World War II interrupted all work that was not related to defense.

When peace came Bell Labs decided to resume their work on these intriguing crystals. Three men, destined to become world famous for their discovery, worked together on what we would now call the Transistor Project: William Shockley, Walter H. Brattain, and John Bardeen.

On December 23, 1947, they were working on a germanium crystal, which is also a semiconductor. In the course of their experiments, they placed two tiny wires 2/1,000 of an inch apart so as to make contact with the surface of the crystal. To their surprise, a telephone voice was amplified 40 times. This unusual ability of a crystal to do what only vacuum tubes had done before was called the *transistor effect*.

For their discovery, the trio received the 1956 Nobel prize in Physics.

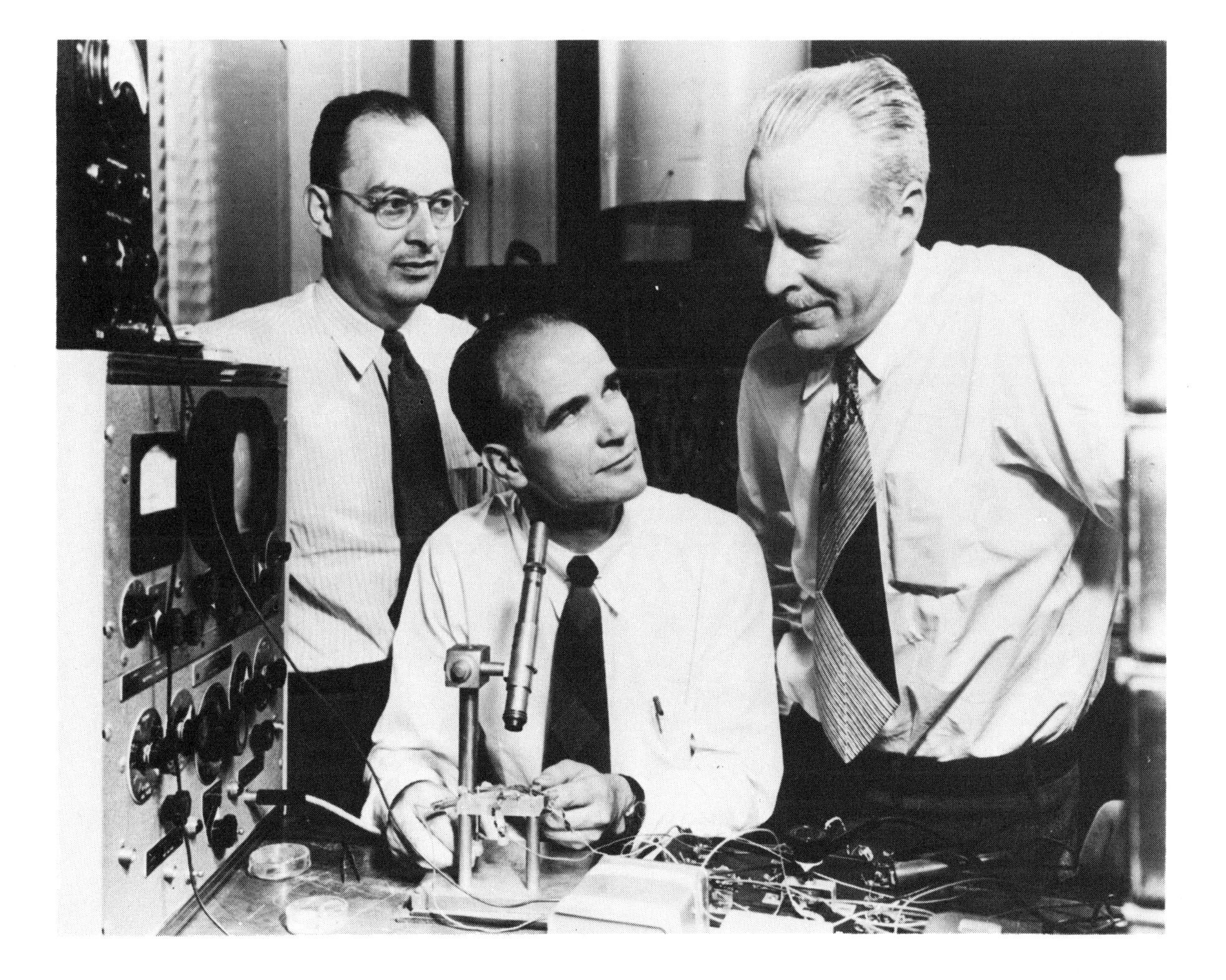

Nobel Prize winners Drs. John Bardeen, William Shockley, and Walter H. Brattain, shown *(left to right)* at Bell Telephone Laboratories in 1948 with the apparatus used in the first investigations that led to the invention of the transistor. The trio received the 1956 Nobel Physics award for their invention of the transistor, which was announced by Bell Laboratories in 1948. (Courtesy Bell Labs)

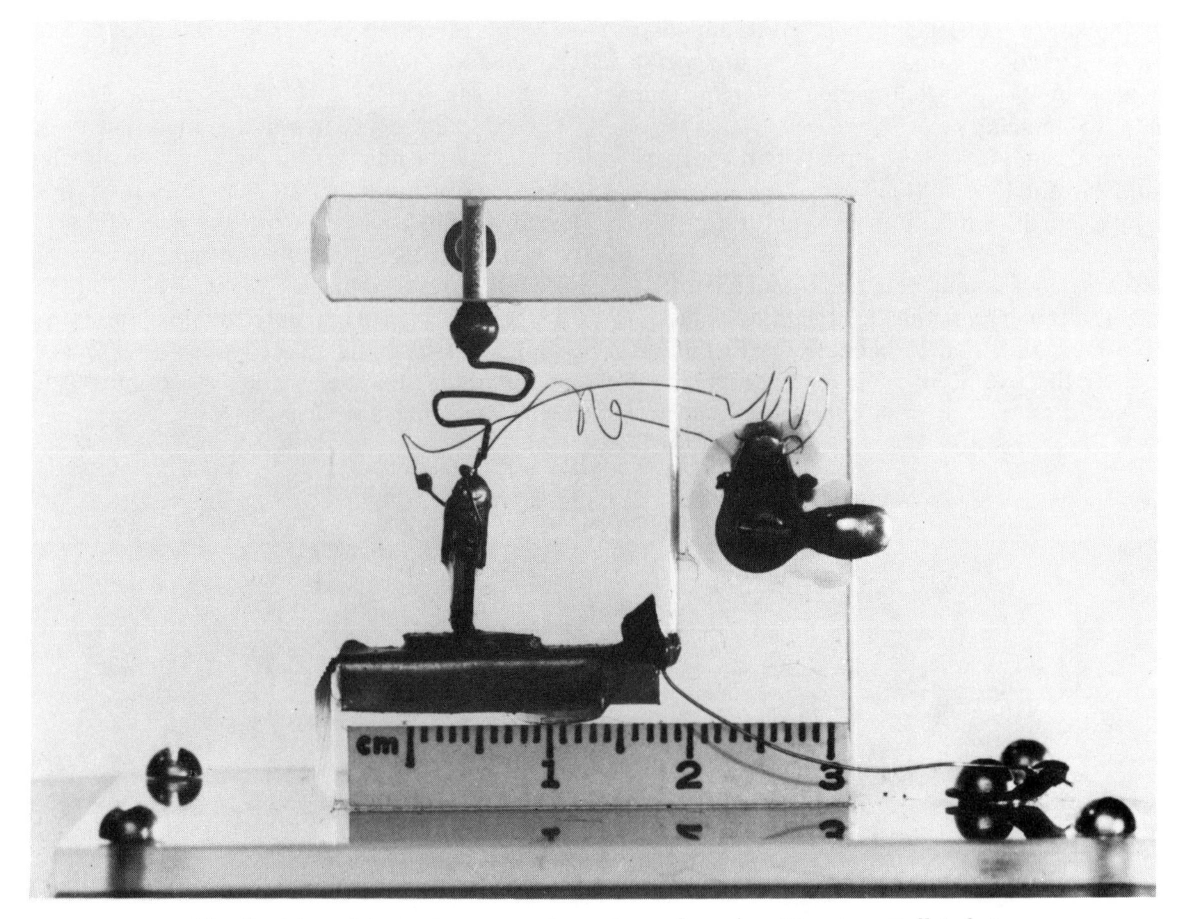

The first transistor—size comparison shown by ruler. (Courtesy Bell Labs)

The Transistor Age

There were still many problems to be solved. In June of 1948, two wires were placed so as to contact a germanium crystal. A third was connected to a battery, providing what is called a *bias current*. It is this bias that transforms a crystal into a conductor.

In England vacuum tubes were called valves, an accurate name since the tube acts as a valve controlling a flow of electrons. The transistor behaves the same way. One wire conducts a tiny signal into the crystal. The amplified signal is led out by a second wire. A third wire is the valve controlling the flow. A small change in the incoming signal makes a big change in the outgoing signal.

The early transistors were noisy and mechanically weak. The lead wires would break off so easily that it was difficult to touch them. Another problem was that there were unwanted impurities that made uniformity impossible. It was hard to tell just how a particular transistor might behave.

A new way of refining the germanium was discovered in 1954. That improved the transistors' behavior, but production was not possible until 1955, the beginning of the transistor age.

One of the greatest values of the transistor is that it takes a tiny signal, so faint the sharpest ear cannot hear it, and amplifies the signal so it can fill a room with sound.

A transistor can work also as a switch

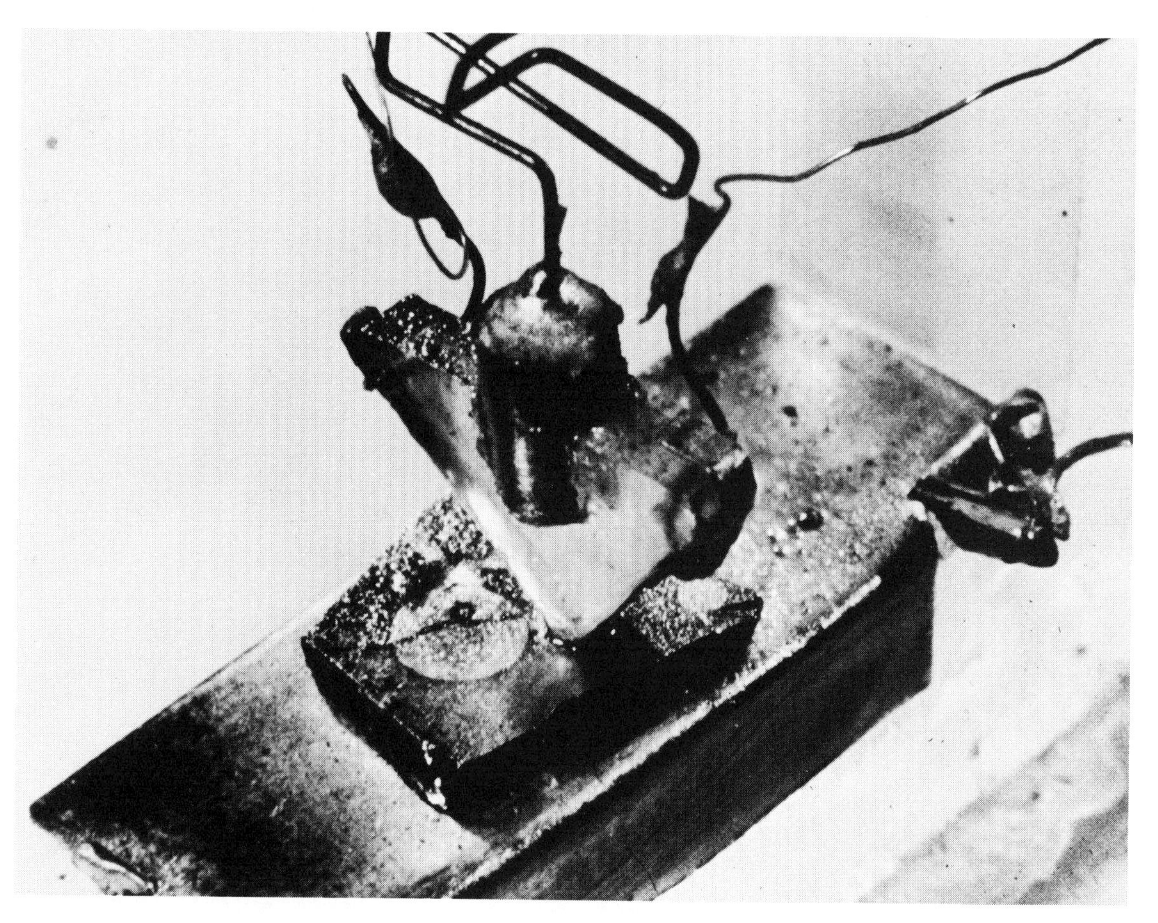

The first transistors assembled by their inventors at Bell Laboratories in 1947 were primitive by today's standards, yet they revolutionized the electronics industry and changed our way of life. The first transistor, a "point-contact" type, amplified electrical signals by passing them through a solid semiconductor material, basically the same operation performed by present "junction" transistors. (Courtesy Bell Labs)

operating at incredible speeds. Computers depend on switches. A computer operating at a few nanoseconds is considered slow! (A nanosecond is one billionth of a second.)

Another difference between the vacuum tube and the transistor is that a tube works in a vacuum. A transistor works in solid material. That's why the term *solid state* is used when referring to transistorized equipment.

Transistors turn on instantly. They require no warm-up time. They run cool and are very small and extremely rugged. Drop a tube and it probably will not work, even if the glass is not shattered. A transistor may operate for years. Little voltage is needed. A portable transistor radio will run for hundreds of hours on one 9-volt battery.

Because of the transistor, equipment has now shrunk dramatically. Years ago, I owned a "portable" tape recorder (using tubes) that must have weighed 30 pounds. Today I can drop a transistorized unit in my shirt pocket and almost forget it's there. It will record for hours while my tube set could record for only about 20 minutes.

The incredible transistor hasn't stopped shrinking yet. Miniaturization has reached a point where thousands of transistors and resistors are packed tightly together in one integrated circuit. A tiny chip may be no bigger than a child's fingernail.

These components are not connected by wires. They are etched together. This makes them very reliable. Their small size allows

The laboratory notebook entry of scientist Walter H. Brattain records the events of December 23, 1947, when the transistor effect was discovered at Bell Laboratories. The notebook entry describes the events and adds: "This circuit was actually spoken over and by switching the device in and out a distinct gain in speech level could be heard and seen on the scope presentation with no noticeable change in quality." (Courtesy Bell Labs)

Comparison of an electron tube with a transistor.

them to be used inside cameras, video games, tape recorders, and medical equipment.

If it were not for transistors inside those tiny chips, the computer that is no bigger than an electric typewriter would not be possible. Without the computers at NASA, space flight would be impossible. And without solar cells, we could not recharge the batteries used to power our space vehicles.

Humanity looked up to the sky in wonderment for thousands of years. Only lately have we been able to escape from the earth's gravity. All this was brought about by the curiosity and imagination of countless experimenters, scientists, and technicians over the centuries in many lands. The study of magnetism, electricity, and electronics brought us step by step closer to the day when we could blast off into the heavens.

Exploring the Tape Recorder

One very modern piece of equipment is a perfect combination of magnetism, electricity, and electronics. It is the tape recorder.

Let's take one apart (in our imagination) to see how it operates. The recording and playback is done on magnetic tape. Such tape consists of tiny, needlelike particles of iron oxide. The iron oxide is coated onto a flexible backing using polyester—the same plastic formula that is used for making clothing. Originally the backing was paper, but plastic stands up better and provides a firmer backing for the iron oxide.

When the tape is blank—with nothing recorded on it—the tiny particles are pointing every which way. When a DC current is applied to the tape as it goes past the erase head, the iron particles are aligned in one direction. Thus the erase head acts as a sort of drum major, forcing every particle into line. The tape is now like a clean blackboard ready for new writing.

The magnetic recording head is made of a special alloy that is easily magnetized when the current is turned on. Turn off the current and the recording head loses its magnetism.

You record by allowing a flow of current to go from the microphone through the coil wound around the head. This creates a magnetic pattern on the tape as it rolls past the head.

The playback head reverses the process. The magnectic field is sensed by the coil around the playback head, and the field creates an electric voltage.

The tape is moved from left to right by means of a motor. Some of the very good home machines and all professional machines have three motors. They are usually AC operated.* The smaller portable units have a single DC motor.

How do you stop the tape? For that matter, how do you start it rolling? Most of the better units employ solenoids to start and stop. This is still another use of a magnetic coil.

Electronics also plays its part in amplifying the sound and playing it back.

From this brief description of a tape recorder, you can see for yourself how magnetism, electricity, and electronics all come together in one machine.

Television

Television, as has been mentioned, uses FM for the audio. The video—the images—is on AM. However, transmitting pictures over the air is much more difficult than transmitting sound. The image is converted into waves from the light and dark areas of a picture. The studio transmits all this information on AM (amplitude modulation), which is a different way of transmitting a signal.

TV is made of very complex circuits made even more complex since color broadcasting began. Soon we will have stereo sound on TV, and even 3D if the scientists working on the idea can make it work.

Exploring Outer Space

Now let's leave earth behind and explore the heavens with a space vehicle, which also depends on electronics to keep it functioning. Once it has been boosted beyond the earth's gravity, a lot of circuits have to be kept operating. They need electricity, but they can't use a

* That is, they draw their current from a wall outlet that is AC. However, this is rectified internally and the machine operates on DC.

For high-quality audible range recording, the iron oxide particles should be longitudinally oriented and uniformly dispersed, as shown in this photograph of a cross-section sample of a 3M Scotch brand magnetic tape. This alignment provides the maximum amount of desirable magnetic behavior of each particle. The sample photographed, obtained by the ultramicrotome slicing method, is eight millionths of an inch thick and magnified by an electron microscope 19,000 times. A common pencil magnified by the same power would be approximately 400 feet in diameter. (Courtesy the 3M Company)

Electron micrograph of 3M magnetic oxide particles. Magnification exceeds 51,000 times. (Courtesy the 3M Company)

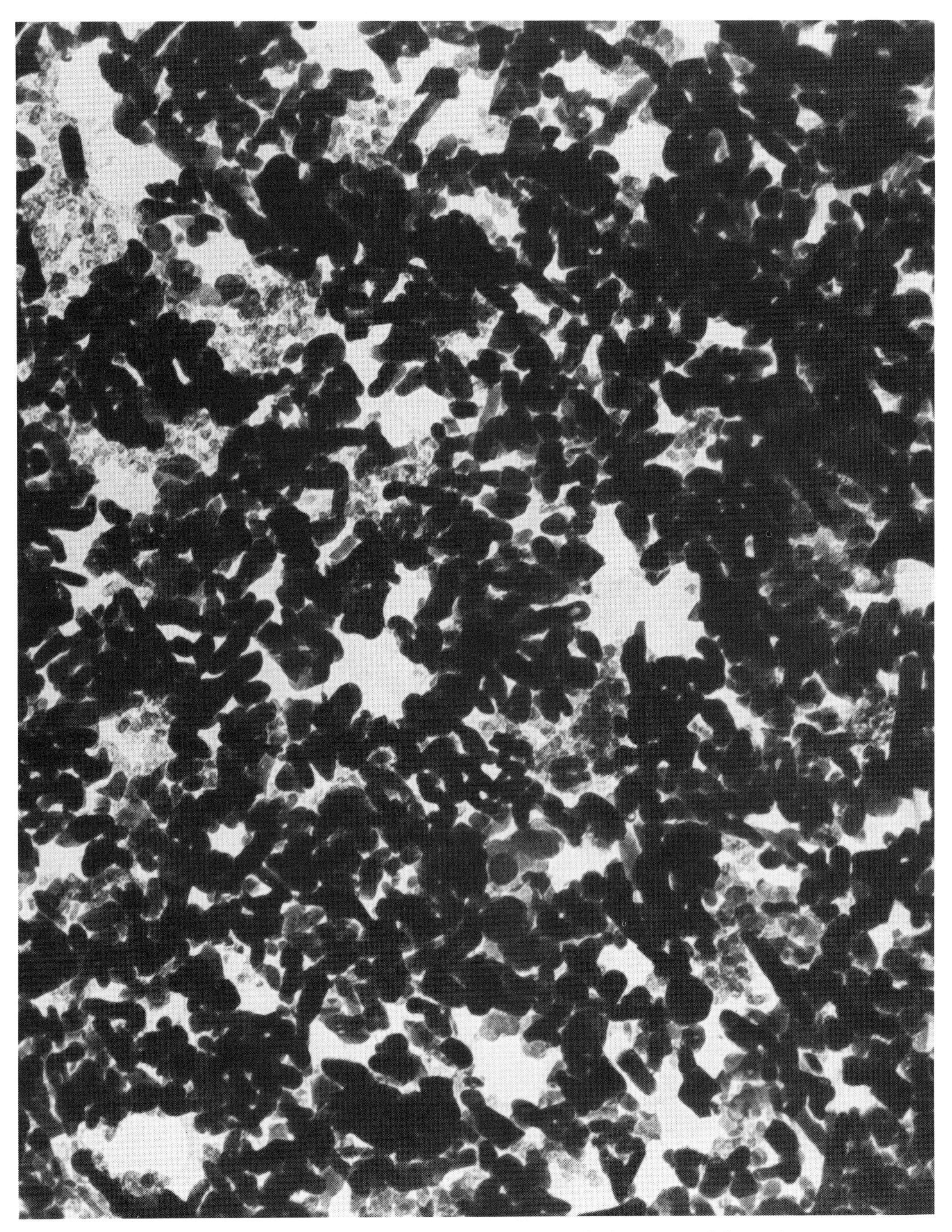

Magnetic tape enlarged some 72,000 times in an electron micrograph shows part of the surface containing the iron oxide particles that are magnetized during the recording process. Oxide particles appear black and the polymeric binder gray. Light areas are voids created in slicing the cross section for viewing. (Courtesy the 3M Company)

generator. It would be too heavy and clumsy. Astronauts need to have their air recycled so they can breathe; they need temperature controls to keep the air at just the right temperature. They need ways to heat the air and ways to cool it whenever necessary, and they need electricity to power their radios and their cameras. And that's only a small part of what a space vehicle requires. Where does it get all that power?

In space, the sun shines all the time. So solar power is the answer. Solar cells are thin sheets made of very pure sand—silicon. These cells convert sunshine into DC electricity and do most of the work for manned or unmanned space probes.

Many electronics dealers sell solar cells. They are not nearly as powerful as those used in space, but then they're not as expensive either. You can buy a kit that includes solar cells and a tiny motor, and some have a radio that operates by solar power. The cell will produce electricity even when the sun doesn't shine. Just use a powerful light. See the list of mail-order dealers on page *viii*, for sources of solar cells and kits. Radio Shack also carries solar equipment.

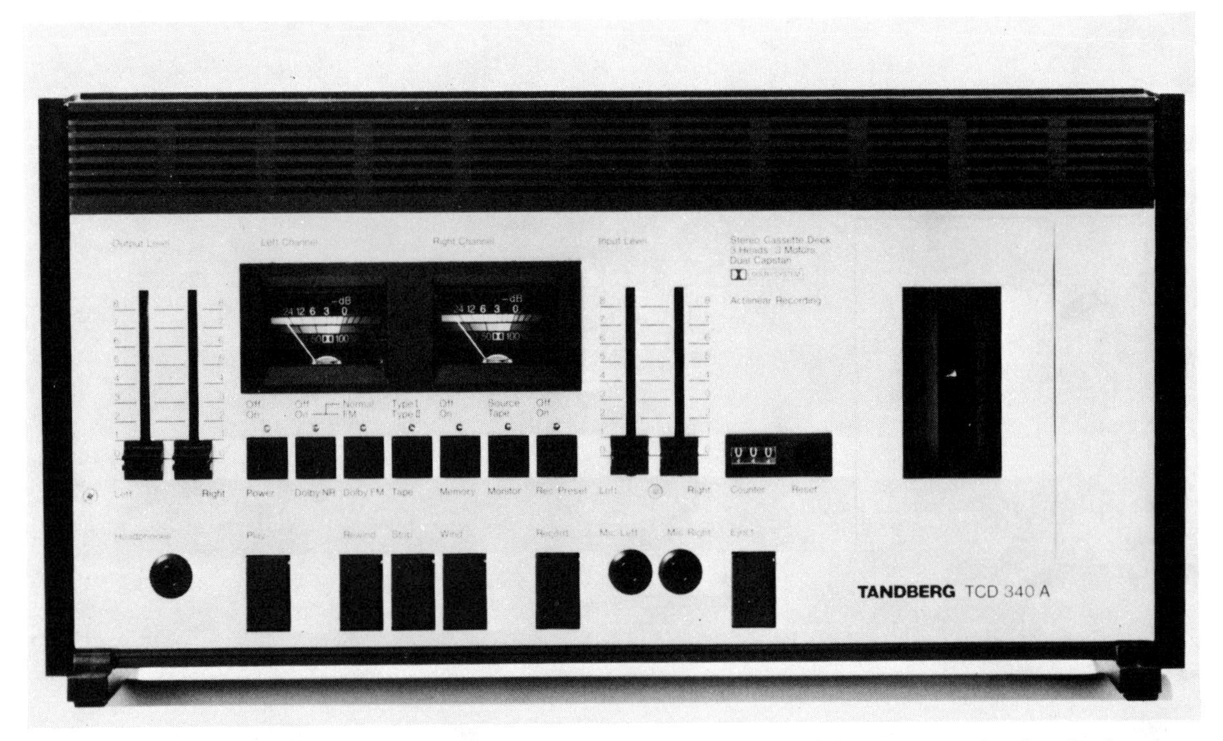

A Tandberg cassette tape player. It plays stereo tracks with three heads and three motors. One head is for playback, one for erasing, and the third for recording. It has volume controls not only for recording but for playback, plus lots of other "goodies." Such a unit combines magnetism, electricity for the motors, and electronics to make the whole thing operative. (Courtesy Tandberg Corporation)

Afterword

The voyage is ended. Together we have traveled through time and through space. Along the way, you met people whose contributions have made our lives better.

Because our journey ends here, does this mean there can be no further progress? Far from it. We are constantly on the edge of great inventions. And the inventors will be among you—the men and women of tomorrow. Even the stars are not the limit.

Bibliography

Aitken, H. G. *Syntony & Spark: The Origins of Radio.* John Wiley & Sons, 1976.
Angell, J. *Elements of Magnetism & Electricity.* Putnam's, 1874.
Atkinson, P. *Elements of Dynamic Electricity & Magnetism.* Van Nostrand, 1891.
Bell, T. H. *Thunderstorm.* Viking Press, 1960.
Benjamin, Park. *History of Electricity.* Arno Press, 1975.
Cheney, M. *Tesla: Man Out of Time.* Prentice-Hall, 1981.
Conot, R. E. *A Streak of Luck (Life of Edison).* Simon & Schuster, 1979.
Gilbert, William. *De Magnete.* Translated by P. F. Mottelay. Dover Publications, 1958.
Hadley, H. E. *Magnetism and Electricity for Beginners.* Macmillan, 1899.
Heilbron, J. L. *Electricity in the 17th and 18th Century.* University of California Press, 1979.
Larsen, Egon. *A History of Invention.* Phoenix House Ltd., 1961.
Li, Dun J. *The Ageless Chinese.* Scribners, 1959.
Marcus and Marcus. *Basic Electricity.* Prentice-Hall, 1969.
Meyer, H. W. *A History of Electricity and Magnetism.* MIT Press, 1971.
Oersted and the Discovery of Magnetism. Blaisdell & Co., 1962.
Roberson, Paul. *What Goes On In Telecommunications?* Warne, Ltd., 1974.
Roller and Roller. *The Development of the Concept of Electric Charge.* Harvard University Press, 1967.
Schaefer, H., ed. *Great Ages of Man—Ancient China.* Time/Life Books, 1967.

Index